Cooling of Electronic Equipment

Cooling of Electronic Equipment

ALLAN W. SCOTT

Varian Associates
Palo Alto, California

A Wiley-Interscience Publication
JOHN WILEY & SONS
New York   London   Sydney   Toronto

Library of Congress Cataloging in Publication Data

Scott, Allan W.
Cooling of electronic equipment.

"A Wiley-Interscience publication."
Includes bibliographical references.

1. Electronic apparatus and appliances—Cooling.
I. Title.
TK7870.S38   621.381   73–11154
ISBN 0–471–76780–8

Printed in the United States of America
10 9 8 7 6 5 4

Preface

This book is written for electronic engineers, electronic technicians, and mechanical engineers who are concerned with the cooling of electronic equipment.

All electronic equipment needs cooling, whether it uses only a few low power transistors or many high power tubes. In most equipment, the cooling design is as important as the electronic design itself.

Unfortunately, no good textbooks are available on the cooling of electronic equipment. The many textbooks on heat transfer and cooling that are available are written for furnace, boiler, and air conditioning designers. These standard textbooks concentrate on the derivation of the basic hydrodynamic equations and on such practical problems as heat exchanger design or the effect of scale in boilers.

The formulas in the standard heat transfer and cooling textbooks are difficult for the electronic engineer to use, because they are expressed in the units of the heat exchanger or furnace designer. It is a formidable task for the electronic engineer to convert from the BTU per hour, °F, and square feet of these formulas to the watts, °C, and inches that are commonly used in designing electronic equipment. Even the tables in the standard textbooks on cooling are useless to the electronic engineer, who has no need to know the thermal conductivity of fire bricks or the thermal properties of carbon tetrachloride.

This book is written to overcome these problems. It is intended for electronic engineers and is written with a terminology and approach they can readily understand.

I am an electronic engineer and my main interest is in electronic design. At first, I tried to ignore the cooling problem, but when some of my finest designs "went up in smoke," I realized I had to face the cooling problem head-on.

My first step at trying to design cooling for electronic equipment was to get out my old college textbooks on hydrodynamics and heat transfer. It was easy enough to understand the theory of thermal radiation, forced convection of air and liquids, or evaporation cooling. Certainly any electronic engineer can understand the concept of Reynolds number, laminar flow, turbulent flow, etc. The problem arises in converting the basic hydrodynamic formulas into a useful form and into an "electronic engineering" system of units.

I soon found that I was not alone among electronic engineers who suffered from a lack of literature on how to cool electronic equipment. This book has grown out of my need to have simple formulas for designing the cooling of my own equipment. Soon I was distributing my notes to fellow engineers, so that they could design the cooling for their equipment. Then I was distributing these notes to our customers, so they could use our equipment without burning it up. I have presented the material of this book in university extension courses and in special industrial courses to electronic engineers who were, out of necessity, also fighting the cooling problem.

This book presents the cooling formulas which can be found in any standard textbook on hydrodynamics or heat transfer in simple form. The formulas are all expressed in terms of power, temperature, dimensions, and the thermal properties of materials. The traditional normalization in terms of Reynolds number, Prandtl number, or the other dimensionless constants of heat transfer theory has been eliminated. An electronic engineer need not learn heat transfer theory, although he probably already knows it, to be able to apply the simple formulas of this book.

Perhaps most importantly, all the formulas in this book are expressed in units of the electronic engineer. Power is in watts, all dimensions are in inches, and temperature is in °C. Thermal properties of all materials commonly used in electronic equipment are tabulated in the same consistent set of units. Air flow and liquid flow are in units which are easily measured in electronic equipment. Finally, a complete appendix is provided with conversion tables to convert from the other systems of units used for heat transfer calculations to the "electronic engineering" systems of units of this book.

A consistent set of symbols has been used throughout the book. All symbols are defined as they are introduced in each chapter, and a summary of symbols is presented in the appendix. Each symbol always represents the same quantity throughout the book, and is always in the same unit.

All the important methods of cooling used in electronic equipment are covered in this book, including:
- Conduction.
- Radiation and natural convection.
- Forced air cooling.
- Forced liquid cooling.
- Liquid evaporation.
- Heat pipes.
- Refrigeration and cryogenic cooling.

Chapter 1 discusses the problem of cooling electronic equipment. The various cooling methods listed above are compared and the design problems that each cooling method presents are described.

Chapters 2 through 6 discuss the basic methods of cooling: conduction, radiation and natural convection, forced air cooling, forced liquid cooling, and liquid evaporation, respectively. Each chapter begins with a discussion of a typical electronic equipment cooled by the method of that chapter. Then each chapter presents the necessary design formulas and the properties of materials commonly used in electronic equipment that are necessary for making calculations with this type of cooling. Where auxiliary equipment is required, such as fans or heat exchangers, available equipment and its capabilities are discussed. At the end of each chapter, sample calculations are given. The first of these sample calculations is usually for the actual electronic equipment used as the illustrative example at the beginning of the chapter. Finally, a list of references is presented in each chapter for further study.

Chapter 7 discusses heat pipes. Chapter 8 describes refrigerated electronic equipment, including cryogenic coolers and thermoelectric cooling modules. Chapter 9 discusses transient cooling effects, such as occur in equipment warmup.

The design information presented in this book gives only approximate answers. Therefore, after the cooling has been designed and the electronic equipment built, the thermal performance of the equipment must be measured. Techniques for making the necessary thermal measurements on electronic equipment are presented in Chapter 10.

It would be nice if we electronic engineers could simply avoid the cooling problem and get on with our electronic design. Unfortunately, the cooling problem won't go away. This book permits the cooling problem to be faced and solved simply and quickly.

I hope it will be of help to you.

Allan W. Scott

Palo Alto, California

Contents

6

7

8

9

Cooling of Electronic Equipment

1

Understanding the Cooling Problem

All electronic equipment needs cooling, whether it uses only a few low power transistors or many high power tubes. In most equipment, the cooling design is as important as the electronic design itself. A typical electronic cooling design is shown in Figure 1.1. A 10 watt transistor is mounted on a finned aluminum heat sink. The transistor is 50% efficient, so 20 W of electrical

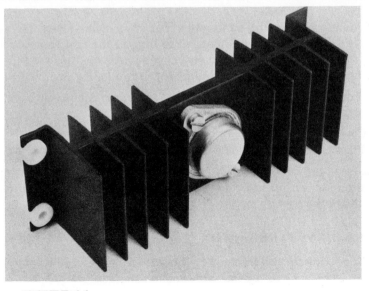

FIGURE 1.1
Transistor mounted on an aluminum heat sink (Photo courtesy of Wakefield Engineering, Inc.)

1

input power are converted to 10 W of useful electrical output power and 10 W of heat. The 10 W of heat must be conducted from the transistor to the cooling fins, where it is then transferred to the surrounding environment.

Heat always flows from hotter objects to cooler ones. Therefore, the transistor must be hotter than the cooling fins, and the cooling fins must be hotter than the surroundings. In any electronic equipment, the temperature of each component will rise until it is hot enough to transfer its internally generated heat to the surroundings. If the cooling design is not adequate, the component will get so hot, in an effort to transfer this heat, that it will destroy itself.

1.1   THE BASIC COOLING PROBLEM

The purpose of cooling electronic equipment is to keep the temperature of each component below its safe operating value.

The electric component temperature is determined from:

$$T \text{ (component)} = \Delta T \text{ (conduction)} + \Delta T \text{ (transfer)} + T \text{ (surroundings)} \quad (1.1)$$

where:

T (component)   is the temperature of the electronic component.

ΔT (conduction) is the temperature difference required to conduct the heat from the electronic component to the cooling fins.

ΔT (transfer)   is the temperature rise of the fins above the surroundings required to transfer the heat from the fins to the surrounding environment.

T (surroundings)   is the temperature of the surrounding environment.

As Equation 1.1 shows, the cooling of electronic equipment consists of two parts:

1. Conduction of heat from the electronic component to the cooling fins.
2. Transfer of heat from the cooling fins to the environment.

This book provides the necessary design information so that the temperature rises required for heat conduction and heat transfer can be calculated, so that the component temperature can be kept within acceptable limits.

1.2   HEAT CONDUCTION

The following factors must be considered in the conduction of heat from the electronic component to the cooling fins:

1. The materials used for conducting heat.
2. The geometry of the heat flow paths.

3. Thermal interfaces where parts are joined together.
4. Conduction of heat through the cooling fins themselves.

Complete design information which considers all the above factors, and which permits a calculation of the conduction temperature rise, ΔT (conduction) of Equation 1.1, is presented in Chapter 2.

1.3   CHOICE OF HEAT TRANSFER METHOD

Once the heat has been conducted from the electronic component to the cooling fins, it must then be transferred to the surrounding environment by one of the following means:

1. Radiation and natural convection.
2. Forced air cooling.
3. Forced liquid cooling.
4. Liquid evaporation.

The above list of heat transfer methods is arranged in order of increasing heat transfer effectiveness. For a given fin area, the least heat can be transferred by radiation and natural convection, more can be transferred by forced air cooling, even more can be transferred by forced liquid cooling, and the most can be transferred by liquid evaporation.

The list is also arranged in order of increasing cooling system complexity. Heat transfer by radiation and natural convection requires no auxiliary equipment—just the cooling fins themselves—and is the simplest design. Forced air cooling requires a fan and fan controls and is more complicated. Forced liquid cooling requires a pump, coolant reservoir, cooling fluid, etc., and is even more complicated.

The effectiveness of each of the above methods of heat transfer is compared in Figure 1.2, which shows the heat that can be transferred per square inch of surface, when the fin surface is at 100°C and the surroundings are at 20°C. For any one heat transfer method the amount of power that can be transferred varies over a wide range, depending on the details of the cooling design.

Figure 1.2 also illustrates that for most applications more than one heat transfer method could be used. The best choice depends on a tradeoff between system simplicity, heat sink complexity, and fin area.

The problems of the design of heat transfer by each of the four methods— radiation and natural convection, forced air cooling, forced liquid cooling, and liquid evaporation—are summarized in the next four sections of this chapter. Complete design information on each heat transfer method is presented in Chapters 3 through 6 respectively. Each of these chapters begins

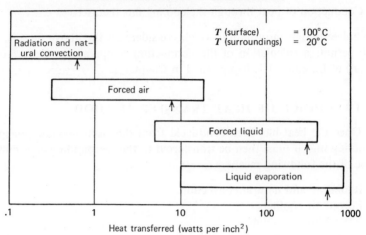

FIGURE 1.2
Relative effectiveness of heat transfer methods

with a typical example of heat transfer in electronic equipment, and the amount of heat transferred per square inch of surface in these examples is shown by the arrows in Figure 1.2.

1.4 RADIATION AND NATURAL CONVECTION

Radiation and natural convection are the simplest to use of all the heat transfer methods. No auxiliary equipment is required, just the cooling fins themselves. The hot fins radiate heat directly to the cooler surroundings. At the same time, the air near the hot fins is heated and rises and is replaced by cooler air. This convective air current provides additional heat transfer.

A typical heat sink cooled by radiation and natural convection was shown in Figure 1.1.

In most electronic equipment, heat transfer occurs by radiation and natural convection simultaneously. However, the amount of heat transferred by each method depends on heat sink temperature, geometry, and orientation in different ways. For a particular electronic cooling design, the heat transferred by each method must be calculated separately and then combined.

The amount of heat that can be transferred by radiation depends on:

1. The temperature of the radiating surface.
2. The temperature of the surroundings.
3. Surface conditions of the fins.
4. Shielding effects of adjacent fins.

The amount of heat that can be transferred by natural convection depends on the following factors:

1. Temperature difference between the surface and the surrounding air.
2. Dimensions of the surface.
3. Orientation of the surface.
4. Spacing between adjacent surfaces.
5. Altitude (which determines air density).

Complete design information on heat transfer by radiation and natural convection, which considers all the above factors, and which permits a calculation of the transfer temperature rise, ΔT (transfer) of Equation 1.1, is presented in Chapter 3.

1.5   FORCED AIR COOLING

As Figure 1.2 shows, the amount of heat that can be transferred from a given cooling fin area is increased by more than an order of magnitude by blowing air over the cooling fins, rather than relying on radiation and natural convection. Forced air cooling is more complicated to implement than cooling by radiation and natural convection, because a fan and the associated fan controls are required. Forced air cooling is, however, much simpler than forced liquid cooling, because a supply of cooling air is readily available and air does not have the freezing, boiling, or dripping problems of cooling liquids.

A typical electronic cooling design using forced air cooling is shown in Figure 1.3. The design of air cooling poses two problems:

1. Choice of the fan or blower.
2. Design of the cooling fin geometry.

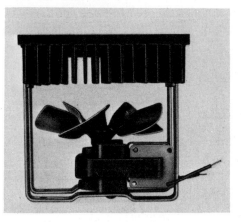

FIGURE 1.3
**Forced air cooled heat sink (Photo
courtesy of Wakefield Engineering)**

These two problems must be solved jointly. The amount of air flow that a particular fan can provide is determined by the pressure into which the fan must work. Both the amount of heat transfer that can be obtained from forced air cooling and the pressure required to force air through the cooling fins depends on air flow and fin geometry. Consequently, the choice of fan must be made in conjunction with the fin design.

Complete design information which permits the most suitable fan and the most optimum cooling fin design to be chosen is presented in Chapter 4. Formulas are presented which permit a calculation of the heat transfer temperature rise, ΔT (transfer) of Equation 1.1, for the case of forced air cooling. The effects of high altitude operation on forced air cooling are included.

1.6 FORCED LIQUID COOLING

Forced air cooling, although simpler to implement than forced liquid cooling, does have several disadvantages. Forced air cooling may not be suitable for electronic equipment which must be operated at high altitudes where air density is low. The acoustic noise of the fan may be objectionable and the vibration of the fan may adversely affect the performance of the electronic equipment. The hot air ejected from the electronic package may also be objectionable.

All of these disadvantages of forced air cooling are eliminated by the use of forced liquid cooling. The cooling pumps and coolant reservoir can be removed from the electronic package, so that quiet, vibration-free operation can be maintained. As shown in Figure 1.2, the use of forced liquid cooling provides an order of magnitude greater heat transfer per unit area than forced air cooling. For very high power electronic equipment, this greater heat transfer capability is a necessity, because the cooling fin area cannot be made great enough to transfer heat by forced air cooling. The use of liquid cooling also permits high density mounting of lower power electronic components. This high density mounting may not only be desirable from a packaging standpoint; it may be a necessity to achieve required electronic performance.

A typical forced liquid cooling design of an electronic component is shown in Figure 1.4. The particular component is the liquid cooled collector of a high power microwave transmitter tube. The coolant inlet and outlet fittings and the multiple liquid cooling ducts can be seen in the cutaway view.

The following factors must be considered in designing forced liquid cooling of electronic equipment:

1. The design of the cooling ducts.

FIGURE 1.4
Forced liquid cooled collector of a high
power microwave transmitter tube
(Photo courtesy of Varian Associates)

2. The type of cooling liquid.
3. The effect of coolant inlet temperature.
4. The coolant pump, heat exchanger, and other auxiliary equipment.

Complete design information for forced liquid cooling, which considers all of the above factors, is given in Chapter 5. Formulas are presented which permit a calculation of the transfer temperature rise, ΔT (transfer) of Equation 1.1, for forced liquid cooling.

1.7   COOLING BY LIQUID EVAPORATION

Evaporation cooling can be effectively used in the following different ways in electronic equipment:

1. Cooling of high power components at high power densities.
2. Cooling of all components in an electronic equipment by immersing the entire assembly in a package filled with dielectric oil.
3. Maintaining a constant temperature bath for electronic components.
4. Simple expendable cooling systems.

Figure 1.5 shows an electron tube cooled by evaporation. The anode of the tube, where heat is generated, is immersed in liquid. Heat is transferred from the tube by boiling the liquid. The vapor from the boiling process rises from

FIGURE 1.5
Cooling of a transmitter tube by liquid evaporation (Photo courtesy of ITT)

the liquid bath, is condensed in a condenser, and is then returned to the cooling bath. The condenser heat transfer area can be made large enough, because it is separated from the electronic equipment, so that heat can be transferred from it to the surroundings by radiation and convection or by forced air cooling.

No coolant pumps are needed for evaporation cooling. Consequently, it is an extremely simple cooling method. As shown in Figure 1.2, evaporation cooling provides greater heat transfer per unit surface area than any of the other methods.

The disadvantages of cooling by evaporation are:

1. The system can operate in one orientation only, with the liquid bath at the lowest point in the system.

2. Immediate destruction of the electronic component occurs if the maximum heat transfer rate is exceeded, because the temperature of the component increases very rapidly above the critical heat transfer rate.

In designing cooling by evaporation, the following factors must be considered:

1. Choice of the cooling fluid.
2. Design of the component surface where evaporation occurs.
3. Design of the condenser.
4. Pressure and pressure equalization.
5. Equipment orientation.

Complete design information for heat transfer by liquid evaporation, including consideration of all the above factors, is presented in Chapter 6.

1.8 HEAT PIPES

The effective cooling design of electronic equipment requires minimizing both the temperature rise required for conducting heat from the electronic component to the cooling fins and the temperature rise required for transferring heat from the cooling fins to the surroundings. As shown in Equation 1.1, both factors contribute to the operating temperature of the electronic component.

In many designs sufficient space may be available for a large number of cooling fins, but the major problem is conducting the heat from the electronic component to the cooling fins. Even if the heat sink is made of copper, the temperature rise due to conduction may be excessive.

Heat pipes offer a solution to this problem. A schematic drawing of a heat pipe is shown in Figure 1.6. The heat pipe consists of a hollow tube which

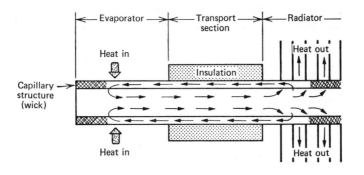

FIGURE 1.6
Schematic drawing of a heat pipe

has been evacuated and then filled with a coolant liquid. The incident heat evaporates the liquid at one end of the pipe and the vapor transports the heat to the other cooler end of the pipe. At the cool end of the pipe the liquid condenses and transfers the heat. So far, this means of heat transfer is the same as evaporation cooling, discussed in the previous section. However, the heat pipe has an additional feature that permits it to be operated in any orientation. The inner surface of the heat pipe contains a capillary structure or "wick" which returns the condensed liquid to the hot evaporator end of the pipe by capillary action. The heat pipe can therefore even operate against gravity, that is, with its evaporator end upward.

For a given temperature rise, a heat pipe can conduct several orders of magnitude more heat than a solid copper rod of the same diameter. Consequently, heat pipes are finding increasing use for the conduction of heat in electronic equipment.

Complete information on the use of heat pipes in electronic equipment is presented in Chapter 7.

1.9   REFRIGERATED EQUIPMENT

The purpose of cooling electronic equipment is to keep the temperature of the electronic components at some desired temperature. With the methods of cooling discussed in all the previous sections, the electronic component is always hotter than the temperature of the surroundings. Heat flows from hot to cold bodies, so the electronic component had to be the hottest element in order to transfer its heat to the surroundings.

If necessary for achieving the desired electronic performance, the electronic component can be cooled to a temperature *below* the temperature of the surroundings by the use of refrigeration. In refrigerated equipment, heat does not flow from the electronic component, but is "pumped" by the refrigeration system from the cold component to the hot surroundings.

Examples of electronic equipment where refrigeration is required are:

- Infra-red detectors.
- Masers.
- Parametric amplifiers.
- Equipment which must work in surroundings which are at a higher temperature than the safe operating temperature of the electronic components.

Refrigeration is not a substitute for good conventional cooling design. If the electronic component can be hotter than the surroundings, then by careful design, the heat can always be transferred from the component by conduction, radiation and natural convection, forced air cooling, forced liquid cooling, or evaporation cooling, and refrigeration is not necessary.

Refrigeration systems for electronic equipment may be classified into the following types:

- Refrigerated cooling air or cooling liquid.
- Refrigerated heat sinks.
- Liquid nitrogen baths.
- Thermoelectric coolers.

These refrigeration systems are discussed in detail in Chapter 8.

1.10 TRANSIENT EFFECTS

Chapters 2 through 8 all consider steady state conditions. Under steady state conditions, heat is transferred to the surroundings as fast as it is generated by the electronic component, and the temperatures of all elements—the component itself, the heat sink, the cooling fins, and the cooling air or liquid—remain constant with time.

However, when electronic equipment is first turned on, or when the power input or the coolant flow rates are changed, the temperature of the electronic component and all other elements in the cooling system vary with time, until the equipment reaches its steady state temperature. Because many electronic components change their electrical characteristics when their temperature changes, the transient thermal characteristics of the equipment must be considered.

Design information on transient thermal conditions in electronic equipment is presented in Chapter 9.

1.11 THERMAL MEASUREMENT TECHNIQUES FOR ELECTRONIC EQUIPMENT

After the cooling of a particular electronic equipment has been designed, and the equipment has been built, the thermal performance of the equipment must be measured. All cooling design formulas give only approximate results, so experimental measurements are essential.

The thermal measurements that must be made on electronic equipment include:

1. Temperature of the electronic component, the heat sink, and the cooling fins.
2. Temperature distribution through the various parts.
3. Air flow rate.

4. Air pressure.
5. Liquid flow rate.
6. Liquid pressure.

Techniques for making the necessary thermal measurements on electronic equipment are presented in Chapter 10.

1.12 SYMBOLS, UNITS, AND CONVERSION FACTORS

A consistent set of units has been used throughout this book. Each symbol is defined as it is introduced in the text, and each symbol stands for the same quantity throughout the book. The units in which each symbol is measured are the same throughout the book. A listing of each symbol and the units in which it is measured is also presented in Appendix A.

This book is intended for use by practicing engineers for the design of cooling of electronic equipment. For this reason the units of measurement used in all formulas have been chosen to be the easiest for electronic equipment designers to use. For example, power or heat is expressed in *watts*, rather than in British thermal units per hour (BTU/hr) or calories per second. This choice of watts as the unit of measurement is obvious, because this is the common measurement unit of power for electronic equipment. All dimensions are expressed in *inches*, rather than in feet or centimeters. Temperature is expressed in degrees Centigrade.

A variety of different systems of units are used in the listed references and other published data. To permit maximum use of other material, a conversion table from all other systems of units to the system of units of this book is presented in Appendix B.

1.13 REFERENCES

Appropriate references are given at the end of each chapter, and the useful information presented in each reference is briefly discussed.

General references on cooling of electronic equipment are as follows:

1. Moore, A. D., *Heat Transfer Notes for Electrical Engineering*, George Wahr Publishing Co., Ann Arbor, Michigan, 1963.

2. *Guide Manual of Cooling Methods for Electronic Equipment*, NAVSHIPS 900, 190, U.S. Government Printing Office, Washington, D.C., 1955.

3. Krauss, A. D., *Cooling Electronic Equipment*, Prentice Hall, Inc., Englewood Cliffs, New Jersey, 1965.

4. Holman, J. P., *Heat Transfer*, McGraw-Hill, New York, 1968.

5. McAdams, W. H., *Heat Transmission*, McGraw-Hill, New York, 1954.

References 1 through 3 describe the cooling of electronic equipment. All contain the basic thermal and hydrodynamic equations from which the formulas, graphs, and tables of this book are derived.

References 4 and 5 are general reference books on heat transmission and cooling.

2

Heat Conduction

The problem of cooling electronic equipment has two parts:

1. Conduction of heat from the electronic component to the cooling fin surface.
2. Transfer of heat from the cooling fins to the surrounding environment.

This chapter discusses the first part of the problem: the conduction of heat from the electronic component, through the heat sink, to the cooling fins. Several factors must be considered:

1. The materials used in the heat sink.
2. The geometry of the heat sink.
3. The mounting interface between the component and the heat sink.
4. Conduction of heat through the cooling fins themselves.

A typical heat conduction design for electronic equipment is described in Section 2.1. The basic heat conduction formula is presented in Section 2.2. The materials of the heat sink and the geometry of the heat sink are discussed in Sections 2.3 and 2.4, respectively. Mounting interfaces are discussed in Section 2.5, and conduction of heat through the cooling fins is described in Section 2.6. Examples of conduction of heat in electronic equipment are given in Section 2.7 to illustrate the use of the equations, figures, and tables of this chapter.

2.1 A TYPICAL HEAT CONDUCTION DESIGN

A typical heat conduction design for electronic equipment—a transistor mounted on a heat sink—is shown in Figure 2.1. Figure 2.2 shows a cross section drawing of the various parts which make up the heat sink.

Heat is generated in the transistor because, like any electronic component,

14

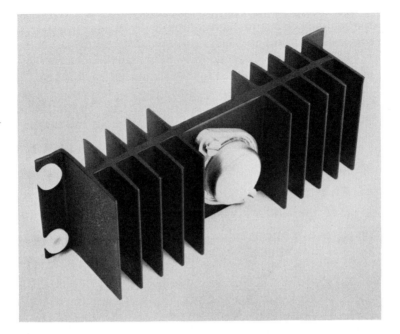

FIGURE 2.1
Transistor mounted on an aluminum heat sink (Photo courtesy of Wakefield Engineering, Inc.)

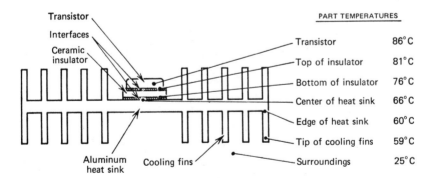

	PART TEMPERATURES	
Transistor	Transistor	86° C
Interfaces	Top of insulator	81° C
Ceramic insulator	Bottom of insulator	76° C
	Center of heat sink	66° C
	Edge of heat sink	60° C
	Tip of cooling fins	59° C
Aluminum heat sink Cooling fins	Surroundings	25° C

FIGURE 2.2
Cross section drawing of the transistor and heat sink of Figure 2.1 showing part temperatures

it is less than 100% efficient. This heat flows across the interface between the transistor and the ceramic insulator, through the ceramic insulator, across the interface between the ceramic insulator and the aluminum heat sink, through the heat sink to the cooling fins, and through the cooling fins themselves where the heat is transferred to the surrounding environment.

Heat always flows from high temperature parts to low temperature ones. Therefore, the electronic component is the hottest part. The total temperature difference required to conduct heat from the electronic component to the cooling fins is given by the following formula:

$$\Delta T \text{ (conduction)} = \Delta T \text{ (interface)} + \Delta T \text{ (insulator)} + \Delta T \text{ (metal)} + \Delta T \text{ (fin)}$$

(2.1)

where:

ΔT (interface) is the temperature difference required to conduct heat across the interfaces where the parts are mounted together.

ΔT (insulator) is the temperature difference required to conduct heat through the insulating elements which make up the heat sink.

ΔT (metal) is the temperature difference required to conduct heat through the metal elements which make up the heat sink.

ΔT (fin) is the temperature difference required to conduct heat through the fins from their base to their tip.

Formulas for calculating all the temperature differences in Equation 2.1 are presented in this chapter.

The fins must be hotter than the surroundings to transfer heat. The temperature difference between the fins and the surroundings depends on the method of heat transfer, whether it is by radiation and natural convection, forced air cooling, forced liquid cooling, or liquid evaporation. This transfer temperature difference is discussed in detail in Chapters 3 through 6.

Figure 2.2 also shows the temperatures of the individual parts of the heat sink. When 10 W of heat are generated in the transistor, it must be 5°C hotter than the top side of the ceramic insulator to conduct the heat through the mounting interface between it and the insulator. The top surface of the ceramic insulator must be 5°C hotter than its bottom surface to conduct the heat through it. The bottom surface of the ceramic insulator must be 10°C hotter than the region of the heat sink directly under it to conduct the heat across this mounting interface. The region of the heat sink directly under the transistor must be 6°C hotter than the edges of the heat sink to conduct heat from the center of the heat sink out to its edges. Finally, the base of the fins must be 1°C hotter than their tip to conduct the heat along the fins. The calculation of these temperature differences, using the design information of this chapter, is illustrated in the first example of Section 2.7.

The total temperature difference from the transistor to the base of the fins must be 26°C to conduct 10 W of heat. If more power must be conducted, the temperature difference will be greater; if less power is conducted, the temperature difference will be less. The temperature difference is approximately directly proportional to the power that must be conducted. Therefore, it is possible to define a thermal resistance for the heat sink, which is as follows:

$$\text{Thermal resistance} \equiv \theta = \frac{\Delta T}{Q} \tag{2.2}$$

where:

ΔT is the temperature difference between the component and the cooling fins (°C).

Q is the power that must be conducted from the component to the fins (watts).

The thermal resistance of the design shown in Figure 2.1 is 2.6°C/W. This means that if 1 W of power must be conducted, the component will be 2.6°C hotter than the cooling fins; for 10 W of power, the component will be 26°C hotter; for 20 W of power, the component will be 52°C hotter, etc.

Because most electronic components, like the transistor, can operate over a range of power, the concept of thermal resistance is extremely useful. It permits the component temperature to be determined at whatever power the component is operating. The thermal resistance of most semiconductor diodes, SCRs, and transistors, is given on the component's data sheet. An example showing the calculation of the junction to case thermal resistance of a high frequency transistor is presented in the second example of Section 2.7 to illustrate the use of the design information of this chapter.

2.2 THE BASIC HEAT CONDUCTION FORMULA

The basic formula describing the conduction of heat through metals and through insulators is:

$$\Delta T = \frac{Q\lambda}{k\alpha} \tag{2.3}$$

where:

ΔT is the temperature difference required to conduct the heat (°C).
Q is the heat that is being conducted (watts).
λ is the length through which heat is conducted (inch).

α is the cross section of the metal or insulator through which the heat is being conducted (inch^2).

k is the thermal conductivity of the metal or insulator (watts/inch °C).

The relationship between the dimensions used in Equation 2.3 is shown in Figure 2.3. An examination of Equation 2.3 reveals that the temperature difference is directly dependent on the power that must be conducted and on

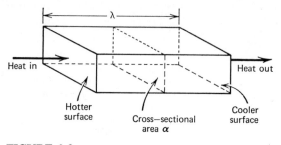

FIGURE 2.3
Geometry for basic heat conduction formula

the length through which it is conducted. The longer the length, the greater the temperature difference. The temperature difference depends inversely on the cross sectional area and on the thermal conductivity of the material. The greater the cross sectional area, the less the temperature difference.

2.3   THE MATERIALS OF THE HEAT SINK

The thermal conductivity of the materials commonly used in electronic equipment is tabulated in Table 2.1. The materials are listed alphabetically in the categories of metals, semiconductors, and insulators.

TABLE 2.1
Thermal Conductivity of Materials Commonly Used in Electronic Equipment

Material	Thermal Conductivity (Watts/Inch °C) at 100°C
Metals	
Aluminum	5.5
Beryllium	4.5
Beryllium copper	2.7
Brass (70% copper—30% zinc)	3.1
Copper	10.0

Continued

TABLE 2.1—*Continued*

Material	Thermal Conductivity (Watts/Inch °C) at 100°C
Gold	7.4
Iron	1.7
Kovar	0.42
Lead	0.87
Magnesium	4.0
Molybdenum	3.3
Monel	0.50
Nickel	2.3
Silver	10.6
Stainless steel 321	0.37
Stainless steel 410	0.61
Steel, low carbon	1.7
Tin	1.6
Titanium	0.40
Tungsten	5.0
Zinc	2.6
Semiconductors	
Gallium arsenide	1.5
Silicon (pure)	3.7
Silicon (doped to resistivity of .0025 ohm-cm)	2.5
Insulators	
Still air	0.0007
Alumina (99.5%)	0.70
Alumina (85%)	0.30
Beryllia (99.5%)	5.0
Beryllia (97%)	4.0
Beryllia (95%)	3.0
Boron nitride (hot pressed)	1.0
Diamond	16.0
Epoxy	0.005
"Thermally conducting" epoxy	0.02
Glass	0.02
"Heat sink compound" (metal oxide loaded epoxy)	0.01
Mica	0.018
Mylar	0.005
Phenolic	0.005
Silicone Grease	0.005
Silicone Rubber	0.005
Teflon	0.005

The thermal conductivity of most of these materials varies somewhat with temperature. The values of thermal conductivity given in Table 2.1 are all for materials at 100°C.

The thermal conductivity of some of the metals listed in Table 2.1 is shown as a function of temperature in Figure 2.4. A similar graph is shown for ceramic insulators in Figure 2.5. Note that the thermal conductivity of the metals changes slowly with temperature, but that the thermal conductivity of the ceramic insulators changes rapidly. For example, when the

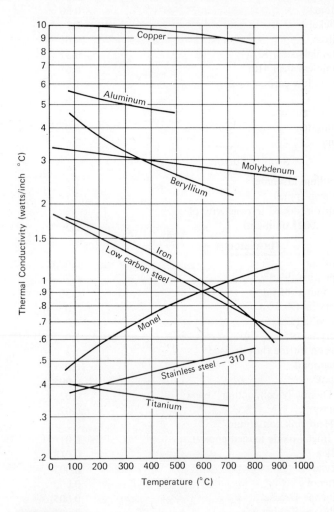

FIGURE 2.4
Thermal conductivity of metals as a function of temperature

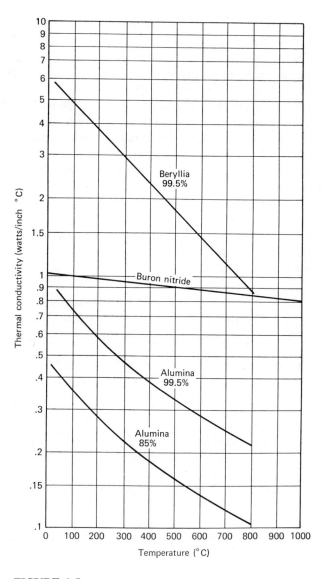

FIGURE 2.5
Thermal conductivity of ceramic insulators as a function of temperature

temperature changes from room temperature to 600°C, the thermal conductivity of the beryllia and alumina ceramics is reduced to one-third of the room temperature value.

The exact value of thermal conductivity of epoxy, phenolic, silicone rubber, and teflon depend on the particular type of material and so may vary around the values given in Table 2.1. The important fact, however, is that all of these materials have extremely poor conductivity, only 1/1000 that of aluminum.

It is instructive to use the basic heat conduction Equation 2.3 and the values of thermal conductivity of various materials from Table 2.1 to develop a feeling for the temperature differences that are to be expected. Assume that a power of 10 W must be conducted through a length of 1 in. in a part with a 1 in^2 cross section. The temperature difference required to conduct this heat through various materials is shown in Table 2.2.

TABLE 2.2
Temperature Difference Required to Conduct 10 W of Heat Through a 1 in. Length of Material with a Cross Section Area of 1 in^2.

Material	Temperature Difference (°C)
Copper	1
Aluminum	2
Low carbon steel	6
Stainless steel #321	25
Beryllia ceramic	2
Alumina ceramic	15
"Thermally conducting" epoxy	500

Of all the commonly used materials, copper is the best thermal conductor. Aluminum is very good also. Stainless steel is poor, with a temperature rise 25 times that of copper. Of the insulating materials, beryllia is extremely good—as good as aluminum metal. Note, however, that the high thermal conductivity of beryllia is critically affected by the material's purity and operating temperature. Alumina ceramic is much poorer than beryllia, but is better than stainless steel.

The poor thermal conductivity of epoxy is clearly evident from Table 2.2. For this comparison, the best thermally conductive epoxy (which has metal particles added to it to improve its conductivity) was used. This material has about four times better conductivity than typical epoxy potting compounds, but even so, the calculated temperature rise is 500°C. In a practical situation the epoxy could not stand this high temperature and would decompose.

2.4   THE GEOMETRY OF THE HEAT SINK

The basic heat conduction formula given in Equation 2.3 applies to designs where the cross section of the conducting metal or insulator is constant over the entire length through which heat is being conducted. Many complicated geometries can be approximately solved by this simple formula by:

1. Assuming an average cross section over the length.
2. Dividing the total length into several segments, each of which has a different cross section, and then applying Equation 2.3 consecutively to each segment.

Two other geometries which are commonly encountered in cooling designs for electronic equipment, and for which simple heat conduction equations are available, are shown in Figures 2.6 and 2.7. Figure 2.6 illustrates radial heat flow in which heat generated at one diameter flows radially outward to another diameter. In this case the cross sectional area of the material is continually increasing. The high power transmitter tubes shown in Figures 1.4 and 1.5 of Chapter 1 have this heat conduction geometry. This type of

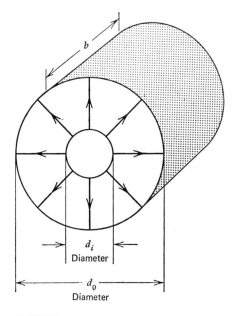

FIGURE 2.6
Radial heat flow geometry

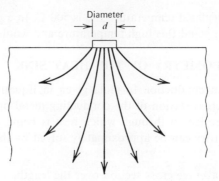

FIGURE 2.7
Spreading heat flow geometry

heat flow geometry also approximates the case of a small electronic component mounted in the center of a large thin heat sink plate.

Figure 2.7 illustrates the case of spreading heat flow where a small electronic component is mounted on an extremely large heat sink. In this case the heat flows radially away from the component as it spreads through the heat sink. The major temperature drop occurs very near the component itself where the heat is confined within a small volume.

The temperature difference required to conduct heat through the radial heat flow geometry shown in Figure 2.6 is given by the following formula:

$$\Delta T = \frac{Q}{2\pi k b} \log \frac{d_o}{d_i} \qquad (2.4)$$

where:

ΔT is the temperature rise required to conduct heat from one diameter to the other (°C).

Q is the power that is being conducted (watts).

d_i is the inner diameter (inch).

d_o is the outer diameter (inch).

b is the axial length of the cylinder (inch).

k is the thermal conductivity of the material (watts/inch °C).

$\log d_o/d_i$ is the natural logarithm of the ratio of the diameters.

The temperature difference required to conduct heat in the spreading heat flow geometry shown in Figure 2.7 is given by the following formula:

$$\Delta T = \frac{2}{\pi} \frac{Q}{kd} \qquad (2.5)$$

where:

    ΔT is the temperature difference between the electronic component and the distant end of the heat sink (°C).

    Q is the power that is being conducted (watts).

    d is the diameter of the electronic component (inch).

    k is the thermal conductivity of the heat sink material (watts/inch °C).

Involved mathematical techniques are described in Reference 1 for the solution of heat conduction in complicated geometries. For most electronic cooling designs, the most practical approach is to synthesize the heat flow geometry into one of the three cases for which simple formulas are available and then make an approximate calculation. When the electronic equipment has been built, the temperature distribution can be measured. Techniques for measuring temperature distribution are presented in detail in Chapter 10.

2.5 MOUNTING INTERFACES

The mounting of the electronic component to the heat sink and the mounting of the various parts of the heat sink to each other present critical thermal problems.

These mounting problems are illustrated in Figure 2.8, where magnified views of the interface between two parts are shown. The surfaces of the parts are not perfectly smooth and flat on a microscopic basis as illustrated in

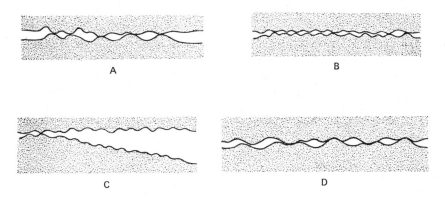

FIGURE 2.8
Magnified views of interfaces between two parts. A. Interface when parts have rough surfaces. B. Interface when parts have smooth surfaces. C. Interface when surfaces are not flat. D. Interface when parts are tightly clamped

Figure 2.8A. Consequently, the two surfaces touch only at their high points. Heat is conducted across the interface only where the surfaces touch. Because they touch at only a few points, the cross sectional area through which heat can be conducted is greatly reduced, often to less than 1% of the total surface area. Unless great care is taken to maximize the contact area, the temperature difference required just to conduct heat across the interface can be as high as that required to conduct heat through several inches of the metal heat sink.

To minimize the temperature difference across the mounting interfaces the following steps can be taken:

1. Improve the surface finish of the heat sink and the electronic component.
2. Apply large clamping pressures to force the component against the heat sink.
3. Solder the electronic component to the heat sink.
4. Use thermal paste ("heat sink compound") between the electronic component and the heat sink.

The Effect of Surface Finish and Contact Pressure

The effect of improving surface finish is illustrated in Figure 2.8B. With a rough surface, such as shown in Figure 2.8A, the parts make contact at only a few points. When the surface finish is improved as in Figure 2.8B, the peaks and valleys along the surfaces are reduced, and the parts make contact at more points. Consequently, the effective cross sectional area for heat conduction is increased.

The overall flatness of the surfaces is even more critical than their surface finish, as illustrated in Figure 2.8C. If the surfaces do not touch at all, no amount of improvement in surface finish will help.

The effect of clamping the surfaces tightly together is illustrated in Figure 2.8D. The microscopic detail of the surfaces is distorted by the clamping pressure to push down the high points and bring the surfaces in better contact. The clamping pressure must be applied uniformly so that the overall flatness of the surfaces is not impaired; otherwise the clamping will bring about the condition illustrated in Figure 2.8C where the surfaces do not touch at all.

Quantitative values for the effect of surface finish and clamping pressure on the temperature difference required to conduct heat across an interface are shown in Figure 2.9. This figure shows the thermal resistance per unit area as a function of contact pressure with surface finish as a parameter. Curves are shown for aluminum and steel heat sinks. Thermal resistance is simply the temperature difference required to conduct heat across the interface per watt of power. As Figure 2.9 shows, by increasing the clamping pressure the temperature difference is decreased. Also, the better the surface finish, the lower the temperature difference at a given clamping pressure.

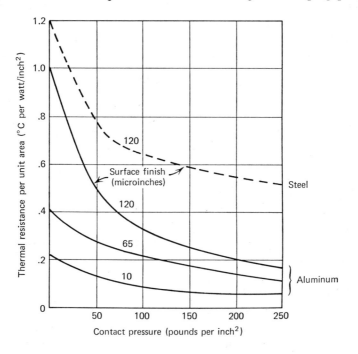

FIGURE 2.9
Thermal resistance of interfaces as a function of contact pressure

Most of the data presented in Figure 2.8 are for two aluminum surfaces. If a different material with a different hardness were used the results would be different, as shown by the one curve for two steel surfaces.

The better the surface finish and the greater the clamping pressure, the greater the amount of heat that can be transferred across the interface for a given temperature rise.

A simple calculation using Figure 2.9 will show the critical nature of the mounting interfaces. Assume that the surfaces of both the electronic component and the heat sink are aluminum and are machined to a 120 microinch finish, that the contact pressure across the surface is 50 psi (pounds per inch^2), that the overall contact area is 1 in^2, and that 10 W of power must be conducted across the interface. From Figure 2.9, the thermal resistance under these conditions is .5°C/W. To conduct 10 W of power across this interface would require a temperature difference of 5°C. Referring back to Table 2.2, this temperature difference is the same as is required to conduct the same 10 W of heat through a 2 1/2 in. length of aluminum bar! By increasing the surface finish to 10 microinches and increasing the contact pressure to 100 psi, the temperature difference across the interface can be reduced to only 1°C for the conduction of 10 W of heat, but this is still as much temperature difference as is necessary to conduct the same amount of heat through a 1/2 in. long aluminum bar.

Soldering the Component to the Heat Sink

The best means of eliminating the mounting interface problem is to solder the electronic component to the heat sink. In this way, the voids between the component and the heat sink surface are filled with solder metal. However, soldering the component to the heat sink does have the following disadvantages:

1. Replacement of the electronic component is difficult.
2. The heat sink and the component must be plated so that the solder will wet the surfaces.
3. Special care must be taken so that the component is not damaged by the heat applied during soldering.

Heat Sink Compounds

An alternative to soldering the component to the heat sink is to clamp the component to the heat sink and use a thermally conductive paste or heat sink compound at the interface. Suitable commercially available heat sink compounds are the following.

Manufacturer	Type
Dow Corning	Type 340 Silicone Heat Sink Compound
Emerson & Cumming	Ecotherm TC-4
Wakefield Engineering	Type 120 Thermal Joint Compound

These heat sink compounds are made of silicones loaded with thermally conductive metal oxides. They are in the form of thick, white pastes which fill the voids between the mating surfaces of the electronic component and the heat sink. They will not leak out of the interface, dry out, harden, or melt even after long exposure at temperatures up to 200°C.

Figure 2.10 shows the correct and incorrect use of heat sink compound. Only a small amount of compound should be applied to just fill the voids

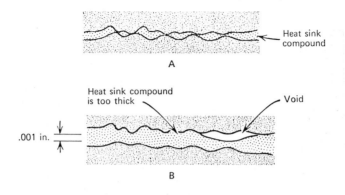

FIGURE 2.10
Magnified views of interfaces when heat sink compound is used.
A. Correct application. B. Incorrect application

and still allow the high points of the metal surfaces to touch, as shown in Figure 2.10A. The surfaces must be perfectly flat so that the thickness of heat sink compound is minimized across the entire surface. Care should be taken so that no brush hairs or foreign particles get into the heat sink compound to separate the surfaces.

Incorrect techniques in using heat sink compound are illustrated in Figure 2.10B.

The heat sink compounds have a thermal conductivity of .01 to .04 W/in. °C. Note that a thermal conductivity of .01 W/in. °C is 1/1000 that of copper. Consequently, the same temperature difference occurs across a .001 in. layer of heat sink compound as occurs across a 1 in. length of copper. For this reason the layer of heat sink compound must be kept very thin.

In spite of their poor thermal conductivity, the use of heat sink compounds can significantly reduce the temperature difference across a mounting interface. A simple calculation will illustrate this. As discussed in a previous paragraph, the thermal resistance across a mounting interface is $.5°C/W/in^2$ if the parts have a 120 microinch surface finish and the clamping pressure is 50 psi. If, instead, the interface is filled with heat sink compound with an average thickness of .001 in., application of Equation 2.3 shows that the thermal resistance is reduced by a factor of 5 to $.1°C/W/in^2$. To obtain this much improvement without the use of heat sink compound would require an improvement in surface finish to 10 microinches and an increase in contact pressure to 100 psi.

Some experimental data which support the above calculation are shown in Figure 2.11. The figure shows the measured thermal resistance between a

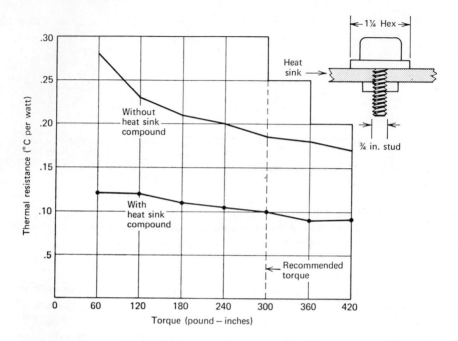

FIGURE 2.11
Measured thermal resistance of the interface between a high power SCR and a heat sink as a function of clamping torque

high power SCR (which has a mounting area of about 1 in²) and its heat sink as a function of the torque applied to clamp the SCR to the heat sink.

The upper curve shows the results when heat sink compound is not used, and as expected the thermal resistance decreases with increasing clamping pressure. The lower curve shows the results when heat sink compound is used. Thermal resistance is reduced by a factor of 2 to .1°C/W, and is relatively independent of clamping pressure.

2.6   CONDUCTION OF HEAT THROUGH THE COOLING FINS

Heat conduction through cooling fins is different than heat conduction through a heat sink because the power that is conducted through the fins is not constant. This effect is illustrated in Figure 2.12A. At any point along the fins, part of the heat is transferred to the surrounding environment and the remainder of the heat is conducted to the adjacent section of the fin. The resulting power at any point along the length of the fin is shown in Figure

2.12B. The power decreases from the total power at the base of the fin to zero at the tip of the fin, since when the tip is reached, all the power has been transferred to the surroundings. The corresponding fin temperature as a function of distance along the fin is shown in Figure 2.12C.

The surface of the cooling fin must be hotter than the surrounding environment to transfer heat. The hotter the fin surface, the greater the amount

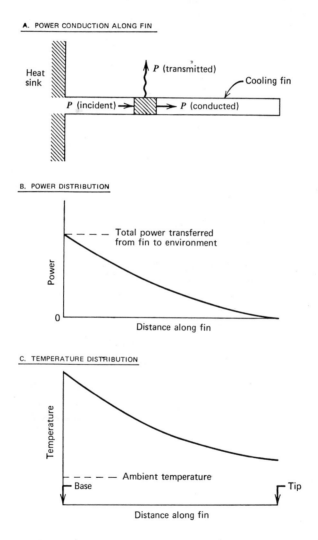

FIGURE 2.12
Conduction of heat through a cooling fin

of heat that can be transferred. Therefore, the temperature difference between the base and the tip of the fin should be minimized, so that the tip of the fin will be as hot as possible to transfer the maximum amount of heat.

"Fin efficiency" is defined as the ratio of the heat actually transferred from the fin to the heat that would have been transferred if the fin surface were at a constant temperature equal to its maximum temperature at the base. Fin efficiency, therefore, takes into account the effect of the temperature difference from the base to the tip of the fin on its heat transfer capability.

Fin efficiency is determined both by the thermal conduction characteristics of the fins and by their heat transfer characteristics. If the heat transferred from the fins is small (for example, if the heat is transferred by radiation and natural convection) then the fins can be long and thin and made of a material whose thermal conductivity is low. If the heat transferred is large (as in the case of forced liquid cooling) then the fins must be thick and short and made of a material with a high thermal conductivity.

Fin efficiency and the base to tip temperature difference are shown in Figure 2.13 as a function of the dimensionless factor:

$$\frac{\left(\dfrac{\Delta T}{Q}\right)_{\text{conduction}}}{\left(\dfrac{\Delta T}{Q}\right)_{\text{transfer}}} \qquad\qquad (2.6)$$

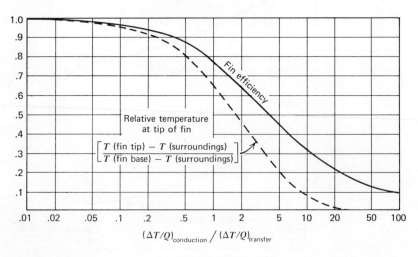

FIGURE 2.13
Fin efficiency and the relative temperature at the tip of the fin as a function of $(\Delta T/Q)_{\text{conduction}}/(\Delta T/Q)_{\text{transfer}}$

where:

$\left(\dfrac{\Delta T}{Q}\right)_{\text{conduction}}$ is the temperature difference required to conduct a constant amount of heat from the base to the tip of the fin, neglecting the effect of heat transfer from the fins.

$\left(\dfrac{\Delta T}{Q}\right)_{\text{transfer}}$ is the uniform temperature difference between the fins and the surroundings required to transfer the constant amount of power from the fins to the surroundings, neglecting the effect of heat conduction through the fins.

The temperature difference required to conduct heat from the base to the tip of the fin, neglecting heat transfer, which is the $(\Delta T/Q)_{\text{conduction}}$ term of Equation 2.6, can be directly calculated from Equation 2.3.

The temperature difference required to transfer this heat from the fins to the surroundings, neglecting the effect of heat conduction, which is the $(\Delta T/Q)_{\text{transfer}}$ term of Equation 2.6, depends on the method of heat transfer used—radiation and natural convection, forced air cooling, or forced liquid cooling. For any given type of heat transfer, the value depends on fin geometry and coolant flow rate, and can be calculated from the design information of Chapters 3 through 6.

Figure 2.13 shows that if the ratio of the temperature difference for heat conduction to the temperature difference for heat transfer is small, which would be the case if the fin were a very good conductor or a very poor radiator, then fin efficiency approaches 100%. If the ratio is one, that is, if the temperature difference required to conduct heat from the base to the tip of the fin is the same as the temperature difference required to transfer the heat from the fin to the surroundings, then fin efficiency is only 76%, and only 76% as much power is transferred from the fin as would be transferred if the entire fin surface were at the temperature of its base. Expressed another way, to transfer the same amount of heat as could be transferred if the fin were a perfect conductor, the base of the fin, where the electronic component is mounted, would have to be $1/.76 = 1.3$ times hotter.

Note also from Figure 2.13, that in this case the temperature difference between the tip of the fins and the surroundings is only 65% of the temperature difference between the base of the fins and the surroundings.

If the ratio of the temperature differences for heat conduction and heat transfer is 20, which would occur if the fin were a very poor conductor or an excellent radiator, then, as shown in Figure 2.13, fin efficiency is only 22% and the temperature of the tip is practically at the temperature of the surroundings. In this case very little power is transferred from the tip end of the fin.

The proper choice of fin material to achieve a high fin efficiency is best illustrated by a simple example. Assume a 1 in. × 1 in. fin whose thickness is 1/16 in. The temperature difference required to conduct a unit of heat from the base to the tip, neglecting heat transfer from the fin, for various fin materials, can be calculated from Equation 2.3, and is as follows.

Fin Material	$\left(\dfrac{\Delta T}{Q}\right)_{conduction}$
Copper	1.6°C/W
Aluminum	2.9°C/W
Steel	9.4°C/W

The temperature difference between the fin surface and the surroundings required to transfer a unit of heat is as follows (from the design information of Chapters 3 through 5).

Heat Transfer Method	$\left(\dfrac{\Delta T}{Q}\right)_{transfer}$
Radiation and natural convection	50°C/W
Forced air cooling	5°C/W
Forced liquid cooling	.5°C/W

Fin efficiency can be determined from the above data for each type of fin material and each cooling method by the use of Figure 2.13. The results are shown in Table 2.3.

TABLE 2.3
Fin Efficiency of a 1 in. × 1 in. × 1/16 in. Fin for Various Methods of Heat Transfer and Various Fin Materials

Heat Transfer Method	Fin Material		
	Copper	Aluminum	Steel
Radiation and natural convection	99%	98%	94%
Forced air cooling	92%	85%	64%
Forced liquid cooling	54%	41%	22%

For a fin with these dimensions, any of the three materials are suitable if the fin is cooled by radiation and natural convection. If forced air cooling is used, copper is the best choice of material, aluminum is satisfactory, and steel should not be used. For the forced liquid cooled case, only copper should be used, and even then the copper fins should be increased in thickness.

2.7 EXAMPLES

The temperature rise of any electronic component above its cooling fins can be calculated from the equations, tables, and figures of this chapter. The correct use of this design information is illustrated in this section by the following examples:

- The transistor heat sink described in Section 2.1.
- A transistor chip inside a stud mounted transistor package.
- A lead mounted silicon diode, potted in epoxy.

EXAMPLE 1
A Transistor Mounted on an Aluminum Heat Sink

Figures 2.1 and 2.2 showed a transistor mounted on an aluminum heat sink Also shown were the temperature differences required to conduct heat across the interface from the transistor to the ceramic insulator, through the ceramic insulator, across the interface between the ceramic insulator and the heat sink, through the heat sink, and finally through the cooling fins themselves from their base to their tip. These temperature differences will be calculated in this first example.

The important dimensions of the transistor and the heat sink are shown in Figure 2.14. The heat that must be conducted from the transistor is 10 W.

The first step is to calculate the temperature difference across the mounting interface between the transistor and the ceramic insulating washer. Heat sink compound .001 in. thick is used at this interface. The temperature difference is calculated using Equation 2.3:

$$\Delta T = \frac{Q\lambda}{k\alpha} = \frac{10 \text{ W} \times .001 \text{ in.}}{.01 \text{ W/in. °C} \times .20 \text{ in}^2} = 5°C$$

where:

λ (the length of heat sink compound through which heat is conducted) is .001 in.

α (the area of the mounting surface of the transistor) is .20 in^2.

k (the thermal conductivity of the heat sink compound) is .01 W/in. °C.

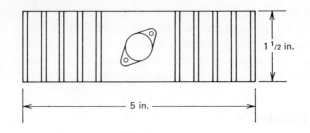

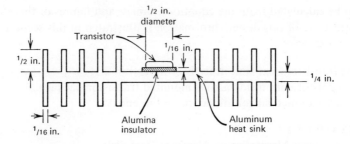

FIGURE 2.14
Dimensions of the transistor heat sink shown in Figure 2.1

The temperature difference required to conduct heat through the heat sink compound at the interface between the transistor and the insulating washer is therefore 5°C.

The second step is to calculate the temperature difference required to conduct the 10 W of heat through the alumina insulating washer. Again Equation 2.3 is used:

$$\Delta T = \frac{Q\lambda}{k\alpha} = \frac{10 \text{ W} \times .063 \text{ in.}}{.7 \text{ W/in. °C} \times .20 \text{ in}^2} = 5°C$$

where:

λ (the length of the alumina washer through which heat is conducted) is .063 in.
α (the cross sectional area of the alumina washer) is .20 in^2.
k (the thermal conductivity of the alumina washer) is .7 W/in. °C.

The value of thermal conductivity of .7 W/in. °C for the alumina ceramic washer was obtained from Table 2.1. The temperature difference required to conduct heat through the alumina washer is 5°C.

The third step is to calculate the temperature difference required to conduct heat across the interface between the alumina ceramic insulator and the heat sink. At this interface heat sink compound is not used. The alumina washer and the heat sink both have a 65 microinch finish and are clamped together with a pressure of 100 psi. The temperature difference required to conduct heat across this interface is determined from Figure 2.9. As shown, for a surface finish of 65 microinches and a clamping pressure of 100 psi, the thermal resistance across the interface is .2°C/W/in². The power density is 10 W/.2 in² or 50 W/in², so the temperature rise across the interface is 10°C. Better design practice would have been to metallize and solder the alumina washer to the heat sink and thereby completely eliminate this element of the total temperature rise.

The fourth step is to calculate the temperature difference required to conduct the heat from the region directly under the ceramic insulator through the heat sink to the base of the fins. The complicated heat flow geometry in this part of the example can be approximated by assuming that half of the power flows in each direction from the centerline of the heat sink to the ends, and using Equation 2.3:

$$\Delta T = \frac{Q\lambda}{k\alpha} = \frac{5 \text{ W} \times 2.5 \text{ in.}}{5.5 \text{ W/in. °C} \times .25 \text{ in.} \times 1.5 \text{ in.}}$$

$$= 6°C$$

where:

λ (the length of the heat sink from the center-line to the edge) is 2.5 in.
α (the cross sectional area of the heat sink) is .25 × 1.5 in².
k (the conductivity of aluminum from Table 2.1) is 5.5 W/in. °C.

The temperature rise from the region directly under the SCR to the outer edge of the heat sink is 6°C.

The final step is to calculate the fin efficiency and the temperature difference from the base of the fins to their tip. Anticipating the results of Chapter 3, $(\Delta T/Q)_{transfer} = 3.5°C/W$ for this cooling fin assembly of 20 fins, and $(\Delta T/Q)_{transfer} = 70°C/W$ for a single fin.
$(\Delta T/Q)_{conduction}$ can be calculated from Equation 2.3:

$$\left(\frac{\Delta T}{Q}\right)_{conduction} = \frac{\lambda}{k\alpha} = \frac{.5 \text{ in.}}{5.5 \text{ W/in. °C} \times .063 \text{ in.} \times 1.5 \text{ in.}}$$

$$= 1°C/W$$

where:

λ (the length of the fin) is .5 in.
α (the cross sectional area of the fin) is .063 × 1.5 in.
k (the conductivity of aluminum) is 5.5 W/in. °C.

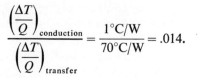

$$\frac{\left(\dfrac{\Delta T}{Q}\right)_{conduction}}{\left(\dfrac{\Delta T}{Q}\right)_{transfer}} = \frac{1°C/W}{70°C/W} = .014.$$

From Figure 2.13, fin efficiency is close to 100% and the temperature difference between the tip of the fins and the surroundings is 99% of the temperature difference between the base of the fins and the surroundings.

Since $(\Delta T/Q)_{transfer}$ is 3.5°C/W for the complete finned heat sink and 10 W of power is being transferred, the base of the fins is 35°C above the surroundings. If the surroundings are at 25°C, the base of the cooling fins are at 60°C. The tip of the fins are 99% × 35°C = 34.6°C above the surroundings or less than 1°C different in temperature from the fin base.

From the above calculations, the total temperature difference from the transistor to the base of the cooling fins is:

	Temperature Difference
Across the interface between the transistor and the insulating ceramic washer	5°C
Through the insulating ceramic washer	5°C
Across the interface between the ceramic washer and the heat sink	10°C
Through the heat sink	6°C
TOTAL	26°C

Since the base of the cooling fins are at 60°C, the transistor mounting surface is at 86°C when 10 W is being dissipated.

EXAMPLE 2
A Transistor Chip Inside a Stud-Mounted Transistor Package

Figure 2.15 shows a photograph of a microwave transistor, and an enlarged drawing of the inside of the transistor showing the silicon chip, that beryllia ceramic insulator, and part of the copper mounting stud. The important dimensions are also shown.

In this example, the junction-to-stud thermal resistance will be calculated. This thermal resistance is the temperature difference per watt of dissipated

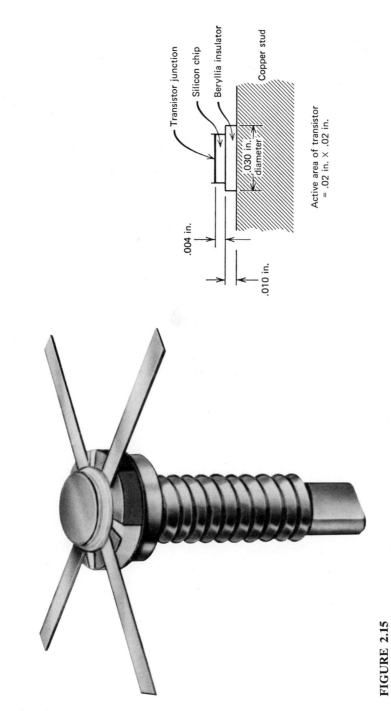

Transistor junction

Silicon chip

Beryllia insulator

Copper stud

.030 in.
diameter

Active area of transistor
= .02 in. × .02 in.

.004 in.

.010 in.

FIGURE 2.15
A high frequency transistor with its important internal dimensions

power between the base-collector junction inside the silicon chip and the mounting stud.

The first step is to calculate the temperature difference required to conduct 1 W of power from the base-collector junction, where the heat is generated, through the silicon chip to the ceramic insulator. This temperature difference can be calculated from Equation 2.3 as follows:

$$\frac{\Delta T}{Q} = \frac{\lambda}{k\alpha} = \frac{.004 \text{ in.}}{2.5 \text{ W/in. }^\circ\text{C} \times .0004 \text{ in}^2} = 4^\circ\text{C/W}$$

where:

λ (the length in the silicon chip through which heat must be conducted) is .004 in.

α (the active area of the transistor) is .0004 in^2.

k (the thermal conductivity of silicon) is 2.5 W/in. $^\circ$C.

The value for the thermal conductivity of silicon was obtained from Table 2.1. The thermal resistance from the junction through the silicon chip is therefore 4°C/W.

The second step is to calculate the temperature difference required to conduct 1 W of power through the beryllia insulator. The silicon chip is soldered to the beryllia insulator and the insulator in turn is soldered to the copper stud so mounting interfaces can be neglected. Again Equation 2.3 is used:

$$\frac{\Delta T}{Q} = \frac{\lambda}{k\alpha} = \frac{.010 \text{ in.}}{4 \text{ W/in. }^\circ\text{C} \times \pi/4 \times (.030)^2 \text{ in}^2} = 3^\circ\text{C/W}$$

where:

λ (the length of the beryllia insulator through which heat is conducted) is .010 in.

α (the cross sectional area of the beryllia insulator) is $(\pi/4)(.030)^2$ in^2.

k (the thermal conductivity of the beryllia insulator) is 4 W/in. $^\circ$C.

The value for the thermal conductivity of beryllia was obtained from Table 2.1. The thermal resistance through the beryllia insulator is therefore 3°C/W.

The third step is to calculate the temperature difference required to conduct 1 W of power through the copper stud from the area directly under the beryllia insulator to the distant end. The "spreading heat flow" formula (Equation 2.5) should be used:

$$\frac{\Delta T}{Q} = \frac{2}{\pi k d} = \frac{2}{\pi \times 10 \text{ W/in. }^\circ\text{C} \times .030 \text{ in.}} = 2^\circ\text{C/W}$$

where:

d (the diameter where the heat enters the copper stud) is .030 in.

k (the thermal conductivity of copper) is 10 W/in. °C.

The thermal resistance to the conduction of heat through the copper stud is 2°C/W.

The final step is to combine the thermal resistance of each part to determine the total junction to stud thermal resistance.

Part	Thermal Resistance (°C/W)
Silicon chip	4
Beryllia insulator	3
Copper stud	2
TOTAL	9°C/W

Therefore, the total junction to case thermal resistance is 9°C/W. If 10 W of power is dissipated in the transistor, the junction must be 90°C hotter than the mounting stud. The mounting stud temperature will be 86°C if the transistor is mounted to the heat sink of Example 1, and in this case the junction temperature would be 176°C.

EXAMPLE 3
A Lead-Mounted Silicon Diode, Potted in Epoxy

All electronic components need cooling, including those which must dissipate only a small amount of power. A typical low power example is a silicon diode rectifier. The total power dissipated in the rectifier is 1 W. Often the electronic equipment will be potted in epoxy, and heat can be transferred from the diode by conduction through its leads and through the epoxy.

A drawing of the silicon diode rectifier with its important dimensions is shown in Figure 2.16.

The first step is to calculate the temperature difference if heat is conducted from the diode through its leads, neglecting conduction through the epoxy.

One-half watt of heat must be conducted through each lead. From Equation 2.3:

$$\Delta T = \frac{Q\lambda}{k\alpha} = \frac{1/2 \text{ W} \times 1 \text{ in.}}{10 \text{ W/in. °C} \times \pi \times (.025)^2 \text{ in}^2} = 26°C$$

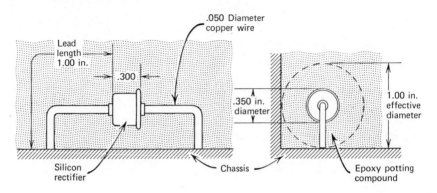

FIGURE 2.16
A lead-mounted silicon diode, potted in epoxy, with its important dimensions

where:

Q (the power that must be conducted through each lead) is 1/2 W.
λ (the length of the leads) is 1 in.
k (the thermal conductivity of copper) is 10 W/in. °C.
α (the cross sectional area of the leads) is $\pi \times (.025)^2$ in².

The temperature difference between the case of the silicon diode rectifier and the heat sink is 26°C.

For comparison, the temperature difference if heat is conducted through the epoxy, neglecting conduction through the leads, should be calculated.

The radial heat flow Equation 2.4 provides a reasonable approximation to this geometry:

$$\Delta T = \frac{Q}{2\pi kb} \log \frac{d_o}{d_i} = \frac{1 \text{ W}}{2\pi \times .005 \text{ W/in. °C} \times .300 \text{ in.}} \log \frac{1.000}{.350} = 112°C$$

where:

d_i (the inner diameter of the epoxy, which is the same as the diameter of the diode) is .350 in.
d_o (the effective outer diameter of the cylindrical heat flow geometry) is 1.00 in.
b (the axial length of the effective cylinder, which is the same as the length of the diode) is .300 in.
k (the thermal conductivity of the epoxy from Table 2.1) is .005 W/in. °C.

The temperature difference required to conduct heat from the diode through the epoxy is approximately four times greater than the temperature difference required to conduct heat through the diode leads. In other words, potting the diode in epoxy does very little to improve the conduction of heat from the diode.

2.8   REFERENCES

Useful references on heat conduction in electronic equipment are:

1. Krauss, A. D., *Cooling Electronic Equipment*, Prentice Hall, Inc., Englewood Cliffs, N.J., 1965. Chapter 1, pp. 1–29, Chapter 6, pp. 112–127.
2. *Guide Manual of Cooling Methods for Electronic Equipment*, Navships Publication 900, 190, Navy Bureau of Ships, Washington, D.C., 1955, pp. 17–27.
3. Goldman, W., *A Primer on Heat Sinking*, Wakefield Engineering, Inc., Wakefield, Mass., 1966.
4. Gutzwiller, F. W., ed., *SCR Manual*, General Electric Company, Chapter 17, pp. 341–365, Syracuse, N.Y., 1967.
5. Rosebury, F., *Handbook of Electron Tube and Vacuum Techniques*, Addison Wesley Publishing Co., Reading, Mass., 1965.

Reference 1 gives the basic heat conduction equations in Chapter 1, and presents a thorough discussion of fin efficiency in Chapter 6.

Reference 2 gives the basic heat conduction equations and a good discussion of thermal interfaces.

Reference 3 discusses the practical problems of mounting semiconductors to heat sinks.

Reference 4 provides instructions for proper mounting of high power SCRs to heat sinks to minimize thermal interfaces.

Reference 5 gives the thermal conductivities of most of the metals and insulators used in electronic equipment.

3

Radiation and
Natural Convection

Radiation and natural convection are the simplest to use of all the heat transfer methods. No auxiliary equipment is required, just the cooling fins themselves. The hot fins radiate heat directly to the cooler surroundings. At the same time, the air near the hot fins is heated and rises and is replaced by cooler air. This convective air current provides additional heat transfer.

A typical heat sink cooled by radiation and natural convection is described in Section 3.1. In most electronic equipment, heat transfer occurs by radiation and natural convection simultaneously. However, the amount of heat transferred by each method depends on heat sink temperature, geometry, and orientation in different ways.

Design information for calculating heat transfer by radiation will be discussed in Section 3.2. Natural convection will be discussed in Section 3.3. For a particular electronic cooling design, the heat transferred by each method must be calculated separately and then combined.

Examples of electronic equipment cooled by radiation and natural convection are presented in Section 3.4 to illustrate the use of the equations and figures of the chapter.

3.1 A TYPICAL COOLING DESIGN USING RADIATION
AND NATURAL CONVECTION

Figure 3.1 shows a typical heat sink designed for cooling by radiation and natural convection. The electronic component is mounted in the center of the heat sink, and heat is conducted from the component to the cooling

44

FIGURE 3.1
Heat sink cooled by radiation and natural convec-
tion (Photo courtesy of Wakefield Engineering,
Inc.)

fins and is then transferred from the cooling fins to the surroundings by
radiation and natural convection. The conduction of heat from the component to the cooling fins was discussed in detail in Chapter 2.

The measured temperature of the cooling fins of the particular heat sink
shown in Figure 3.1 is shown in Figure 3.2 as a function of the power dissipated by the electronic component. At a power dissipation of 60 W, the
temperature of the fins is 100°C, which is 80°C above the temperature of the
surroundings. Of the total transferred power of 60 W, approximately 20 W
is transferred by radiation and 40 W is transferred by natural convection.

The effectiveness of the various methods of heat transfer were compared
in Figure 1.2 of Chapter 1. The value shown for radiation and natural convection was taken from the cooling fin design shown in Figure 3.1. When the
fin temperature is 100°C and temperature of the surroundings is 20°C, 60 W
of power is transferred. The total surface area of the fins is 107 in², so the
heat transferred is .6 W/in².

The heat transferred from the heat sink shown in Figure 3.1 is calculated
in Example 1 of Section 3.4, to illustrate the design information of this
chapter.

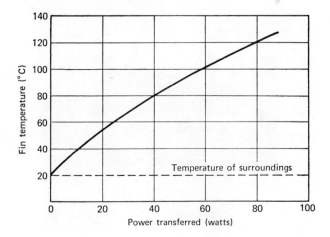

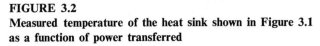

FIGURE 3.2
Measured temperature of the heat sink shown in Figure 3.1 as a function of power transferred

3.2 RADIATION

The amount of heat that can be transferred by radiation depends on:

1. The temperature of the radiating surface.
2. The temperature of the surroundings.
3. Surface conditions of the fins.
4. Shielding effects of adjacent fins.

These factors are related by the following equation:

$$Q = Q_r \varepsilon R S \qquad (3.1)$$

where:

Q is the power transferred by radiation (watts).
Q_r is the power per unit area transferred by a perfect, unshielded radiator, and depends only on the temperature of the surface and the temperature of the surroundings (watts per inch^2).
ε is the emissivity of the surface.
S is the area of the surface (inch^2).
R is the reduction in effective surface area caused by the shielding effects of adjacent surfaces.

Design information on each of the factors in Equation 3.1 is presented in the following paragraphs.

Temperature Dependence

The value of the factor Q_r of Equation 3.1, which is the amount of heat radiated per unit area of surface from a perfect, unshielded radiator, is shown as a function of surface temperature in Figure 3.3. Figure 3.3A shows radiated power over a wide range of surface temperatures up to 700°C. Figure 3.3B shows the same data with an expanded temperature scale over the temperature range up to 350°C. The parameter of the curves is the temperature of the surroundings into which the heat is radiated.

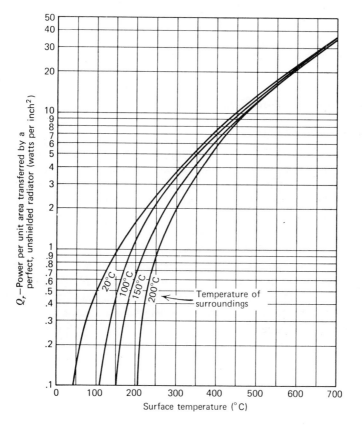

FIGURE 3.3
Power transferred by radiation as a function of surface temperature

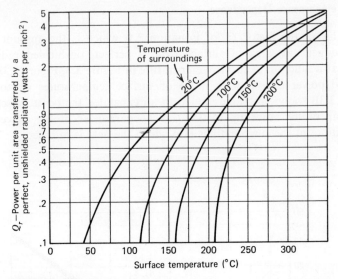

FIGURE 3.3A
Power transferred by radiation as a function of surface temperature

Surface Conditions

The curves of Figure 3.3 show the amount of heat that can be transferred by radiation from a perfect black body radiator. The factor specifying how good a particular surface is relative to a perfect radiator is the "emissivity" of surface, and is the factor ε in Equation 3.1. The emissivity of various materials and various surface treatments commonly used in electronic equipment is shown in Table 3.1.

Table 3.1 clearly shows the importance of surface characteristics on the amount of heat that can be transferred by radiation. If the surface is a polished metal, the heat that can be transferred by radiation is reduced to less than one-tenth of the value shown for an ideal surface in Figure 3.3. Oxidizing the metal surface improves its radiation capabilities to about two-thirds of the values shown for an ideal surface. Note that painting the fin surface with any color paint improves its emissivity to practically the ideal level.

Shielding by Adjacent Fins

The reduction in the effective radiating fin area due to the shielding effect of adjacent fins is shown in Figure 3.4. This is the factor R of Equation 3.1, and it is shown as a function of the ratio of fin height to fin spacing. Separate curves are shown for a square fin whose height is equal to its length, for a rectangular fin which is twice as long as it is high, and for a long rectangular

TABLE 3.1
Emissivity of Materials Commonly Used in
Electronic Equipment

Surface	Emissivity
Commercial aluminum (polished)	.05
Anodized aluminum	.80
Aluminum paint	.27 to .67
Commercial copper (polished)	.07
Oxidized copper	.70
Stainless steel (polished)	.17
Stainless steel (with heavy oxide)	.85
Rolled sheet steel	.66
Air drying enamel (any color)	.85
Oil paints (any color)	.92
Lamp black in shellac	.95
Varnish	.90
Zirconium coating on molybdenum	.65

fin whose length is very much greater than its height. The definitions of the width, the length, and the spacing of the fins are shown in the inset of Figure 3.4.

The shielding effect of adjacent fins significantly reduces the effective radiating area of a finned surface.

A simple calculation using Figure 3.4 clearly illustrates this. Assume heat is to be radiated from a 2 in. × 2 in. surface at 100°C. The surroundings are 20°C, and the surface has been painted black to maximize emissivity. From Figure 3.3, this surface will radiate .5 W/in^2 × 4 in^2 = 2.0 W. Next, assume that the surface is finned with 1 in. high by 2 in. long fins so that it appears as shown in the inset of Figure 3.4. Table 3.2 shows the amount of heat that will be radiated for different numbers of fins, using the curves of Figure 3.4.

The outer surface of the end fins are unshielded and have the same area as the original surface. The inner surface of the end fins, all other fins, and the original surface are all shielded by each other and their effective radiating area is reduced. Adding two fins at each end of the original surface increases the total area by 3 times. The shielding factor is .86 for this case, and the total radiated power is increased by 2.5 times from the case of the unfinned surface. Adding more fins in addition to the two end fins provides very little improvement. For example, when 9 fins are used, the total surface area is increased more than 3 times from the two fin case, but the shielding factor is .3, so the radiated power is increased only 1.35 times.

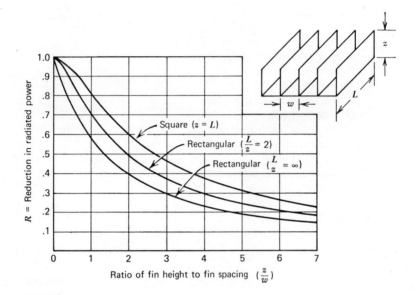

FIGURE 3.4
Radiation shielding effect of adjacent surfaces

TABLE 3.2
Radiation From a Finned Surface

Number of Fins	Fin Spacing (inch)	Added Area (inch²)	Total Area (inch²)	$\frac{z}{w}$	R	Effective Added Area (inch²)	Radiated Power (watts)
0	—	0	4	—	—	—	2.0
2	2	8	12	1/2	.86	6.9	5.5
3	1	12	16	1	.72	8.6	6.3
5	1/2	20	24	2	.50	10.0	7.0
9	1/4	36	40	4	.30	10.8	7.4

3.3   NATURAL CONVECTION

Figure 3.5 shows sketches of natural convection air flow over various orientations of heated surfaces. In the vertical orientation shown in Figure 3.5A, which is one of the best orientations for natural convection cooling, the hot air rises from the surface and is replaced by cool air from the bottom. In the case of a horizontal surface facing upward shown in Figure 3.5B, cool air flows along the surface from the sides and as it is heated, rises from the surface. If the horizontal surface faces downward, as shown in Figure 3.5C,

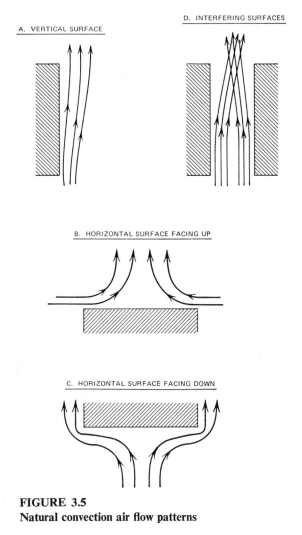

FIGURE 3.5
Natural convection air flow patterns

the heated air must flow along the surface to the edge before it can then rise, and for this reason, convective cooling of this orientation is only half as effective as if the surface were vertical or facing upward.

Figure 3.5D shows two closely spaced vertical surfaces, where the heated air mixes near the top of the surfaces.

The sketches of Figure 3.5 suggest that the amount of heat that can be transferred by natural convection depends on the following factors:

1. Temperature difference between the surface and the surrounding air.
2. Dimensions of the surface.
3. Orientation of the surface.
4. Spacing between adjacent surfaces.
5. Altitude (which determines air density).

These factors can be related by the following equation:

$$Q = Q_c R_1 R_2 S \qquad (3.2)$$

where:

Q is the power transferred by natural convection (watts).

Q_c is the power per unit area that is transferred from a 1 in. long vertical surface, and depends only on the temperature difference between the surface and the surroundings (watts per inch^2).

R_1 is a factor which depends on surface dimensions and orientation.

R_2 is a factor that depends on altitude.

S is the area of the surface (inch^2).

Design information on each of the factors in Equation 3.2 is presented in the following paragraphs.

Temperature Dependence

Figure 3.6 shows the power that is transferred by natural convection from a 1 in. high vertical surface as a function of the temperature difference between the surface and the surroundings. This is the factor Q_c of Equation 3.2.

If the surrounding air is at 20°C, and the surface is at 100°C, so that the temperature difference is 80°C, then as Figure 3.6 shows, 1/2 W of power is transferred by natural convection per inch^2 of surface area. This is the same amount of power as is transferred by radiation from a perfect, unshielded radiator at these same surface and surrounding temperatures.

Surface Dimensions and Orientation

The airflow sketches of Figure 3.5 suggest that the effectiveness of natural convection cooling depends on the length of the surface over which air moves and on the orientation of the surface. Figure 3.7 shows the factor R_1 of

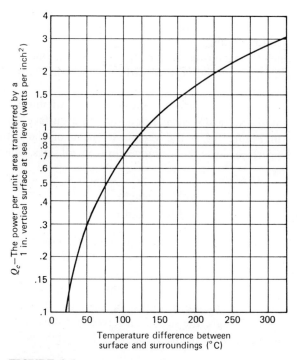

FIGURE 3.6
Power transferred by natural convection as a function of the temperature difference between the surface and the surroundings

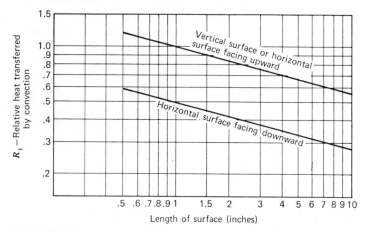

FIGURE 3.7
Relative effectiveness of natural convection cooling as a function of surface dimensions and orientation

Equation 3.2, which includes both of these effects. The relative effectiveness of natural convection cooling (R_1) is shown as a function of the "effective length" over which the air moves. Two curves are shown. One is for horizontal surfaces facing upward and vertical surfaces; the other is for horizontal surfaces facing downward.

The effective length of the surface along which the air moves is equal to the vertical height of a vertical surface. For a horizontal surface, the "effective length for use in Figure 3.7 is the geometric mean of the actual length and width of the surface, since the air moves in from all sides. The geometric mean is given by:

$$\frac{\text{width} \times \text{length}}{\text{width} + \text{length}}$$

For a 1 in. high vertical surface, the relative effectiveness factor (R_1) is 1, so the heat transferred per unit area by natural convection is simply the Q_c term of Equation 3.2, which was shown as a function of temperature in Figure 3.6. For other surface lengths or orientations, the relative effectiveness factor R_1 shown in Figure 3.7 must be included. For example, if a vertical surface is 3 in. high, the heat transferred per unit area by natural convection is reduced to 76% of the value for a 1 in. high surface shown in Figure 3.6.

As expected, cooling by natural convection is much better for a vertical surface or for a horizontal surface facing upward than for a horizontal surface facing downward.

For this reason, a heat sink such as shown in Figure 3.1 should always be operated with its fins vertical rather than horizontal. In the horizontal mounting, as shown by Figure 3.7, the transfer of heat by natural convection from the lower surface of the fins would be reduced by one half.

Spacing Between Adjacent Surfaces

The design information shown in Figures 3.6 and 3.7 applies to unshielded surfaces. If two surfaces are too close together, as shown in Figure 3.5D, the air flow from the surfaces interferes and the design curves no longer apply. For most surface dimensions used in electronic equipment, surfaces or fins can be brought to within 1/2 in. of each other without much mixing of the air currents, and the design information of Figures 3.6 and 3.7 can be used down to this spacing. The effects of spacing between adjacent surfaces is therefore much less critical for natural convection cooling than for radiation cooling.

Altitude

The amount of natural convection cooling that can be obtained depends on air density, and so is a function of the altitude at which the equipment is operated.

Figure 3.8 shows the reduction of convective heat transfer as a function of altitude. This is the factor R_2 of Equation 3.2. As Figure 3.8 shows, convective heat transfer is only 80% as effective at an altitude of 10,000 ft as it is at sea level. At 35,000 ft it is only half as effective.

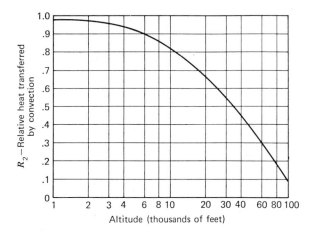

FIGURE 3.8
Relative effectiveness of natural convection cooling as a function of altitude

3.4 EXAMPLES

The amount of heat that can be transferred by radiation and natural convection can be calculated from the equations and figures of this chapter. The correct use of this design information is illustrated in this section by the following examples:

- The finned heat sink described in Section 3.1.
- A radiation cooled high power vacuum rectifier tube.
- A rack of card mounted components.
- Packaged airborne electronics equipment.

EXAMPLE 1
A Finned Heat Sink

A typical heat sink for electronic components, cooled by radiation and natural convection, was described in Section 3.1 and was shown in Figure 3.1.

The power transferred from this heat sink when its fins are at 100°C and the surroundings are at 20°C will be calculated in this example. The important dimensions of the heat sink are shown in Figure 3.9. (The orientation of the heat sink in Figure 3.9 was chosen to best show its dimensions. In actual use, the heat sink would be mounted with the fins vertical.)

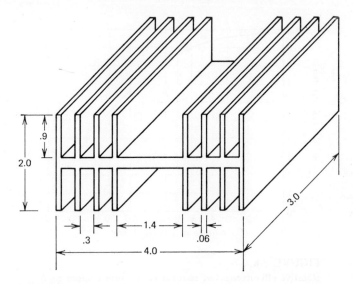

FIGURE 3.9
Dimensions of heat sink shown in Figure 3.1

The first step is to calculate the heat transferred by radiation, using Equation 3.1 and Figures 3.3 and 3.4. From Figure 3.3, when the heat sink surface is at 100°C and the surroundings are at 20°C, the radiated power is .46 W/in². The heat sink is painted black and, using Table 3.1, emissivity (ε) is .95.

The radiating area of the fins consists of the end surfaces which are not shielded and the U-shaped troughs between the fins which are shielded. For the 12 troughs where the fins are spaced .3 in. apart, the ratio of fin height to fin spacing is .9 in./.3 in. = 3 and the shielding factor R from Figure 3.4 is, therefore, .35 for rectangular fins whose length is three times their height.

For the two troughs in the center of the heat sink, the ratio of fin height to fin spacing is .9 in./1.4 in. = .64, and the shielding factor is .8.

The effective radiating area is:

12 troughs × (.9 + .3 + .9) in. × 3 in. × .35
+ 2 troughs × (.9 + 1.4 + .9) in. × 3 in. × .80
+ 2 ends × 2 in. × 3 in. × 1
= 54 in².

The effective radiating is 54 in², whereas the total surface area is 107 in². The power transferred by radiation, using Equation 3.1 is:

$$Q = Q_r \varepsilon R S$$
$$= .46 \text{ W/in}^2 \times .95 \times 54 \text{ in}^2$$
$$= 24 \text{ W.}$$

The second step is to calculate the heat transferred by convection using Equation 3.2 and Figures 3.6 and 3.7.

The temperature difference between the fins and the surroundings is 100°C − 20°C = 80°C. From Figure 3.6, the heat transferred by convection is .53 W/in² for a 1 in. long vertical fin. The heat sink should be mounted with all fins vertical. The effective length along which the air moves is therefore 3 in. From Figure 3.7, the reduction factor R_1, which takes into account fin length and orientation, is .76. Assuming sea level operation, $R_2 = 1$. The spacing between fins of .3 in. is so small that some error will be introduced because of mixing of the air streams from adjacent fins. This error will be ignored in this calculation.

The heat transferred by convection, calculated from Equation 3.2, is therefore:

$$Q = Q_c R_1 R_2 A$$
$$= .53 \text{ W/in}^2 \times .76 \times 1 \times 107 \text{ in}^2$$
$$= 42 \text{ W.}$$

The final step is to combine the effects of radiation and convection to find the total heat transferred from the heat sink.

Radiation	24 W
Convection	42 W
Total	66 W

The above calculations assume a perfectly conducting fin. The fin temperature must be corrected to account for the finite thermal conductivity of the aluminum fin. From Equation 2.3 of Chapter 2:

$$\left(\frac{\Delta T}{Q}\right)_{\text{conduction}} = \frac{z}{kyL} = \frac{.9 \text{ in.}}{5.5 \text{ W/in.°C} \times .06 \text{ in.} \times 3 \text{ in.}} = 0.87°C/W$$

where the thermal conductivity of aluminum from Table 2.1 of Chapter 2 is 5.5 W/in. °C.

$(\Delta T/Q)_{transfer}$ is 80°C/66 W = 1.2°C/W for the heat sink of 16 fins, and, is therefore, 16 times greater or 19°C/W for a single fin.

$$\frac{\left(\dfrac{\Delta T}{Q}\right)_{conduction}}{\left(\dfrac{\Delta T}{Q}\right)_{transfer}} = \frac{.87°C/W}{19°C/W} = .046$$

and from Figure 2.13 of Chapter 2, fin efficiency is 98%. Therefore, the total power transferred from the fins is .98 × 66 W = 65 W.

The measured performance of this heat sink was shown in Figure 3.2. The calculated performance is in excellent agreement with the measured results.

EXAMPLE 2
A Radiation Cooled High Power Vacuum Rectifier Tube

In most electronic cooling designs, the temperature of the cooling fins must be limited to less than 300°C to prevent damage to electronic components and oxidation of the components and the fins. An exception to this rule is the radiation cooling of a high power vacuum rectifier tube. A photograph of such a tube is shown in Figure 3.10.

The power dissipated at the anode of the tube is radiated directly through the glass bulb to the surroundings. The vacuum environment around the anode protects it from oxidation so it can operate "red hot" at 650°C, and take advantage of the high radiation heat transfer that occurs at high temperatures. The anode is a 2 in. diameter × 4 in. long cylinder of molybdenum.

Molybdenum is used for the anode because of its good mechanical and vacuum properties at high temperature. The emissivity of any shiny metal is less than .1, so the outer radiating surface of the anode is coated with zirconium, which, as Table 3.1 shows, raises emissivity to .65.

Only radiation need be considered in this problem. Natural convection does not occur because the anode is in a vacuum.

The amount of heat transferred by radiation is calculated from Equation 3.1. From Figure 3.3, the power radiated from a perfect unshielded radiator is 28 W/in² when the radiating surface is at 650°C, and the surroundings are at 20°C.

Emissivity of the coated molybdenum surface is .65, and R is 1 since the anode is unshielded. The area of the anode is 2 in. × π × 4 in. = 25 in². From Equation 3.1:

$$Q = Q_r \varepsilon R S$$
$$= 28 \text{ W/in}^2 \times .65 \times 1 \times 25 \text{ in}^2$$
$$= 455 \text{ W}.$$

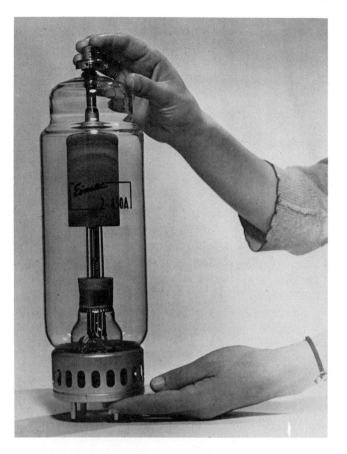

FIGURE 3.10
Radiation cooled high power vacuum rectifier tube (Photo courtesy of Varian Associates)

Therefore, at the rated anode dissipation of the tube of 450 W, the anode operates at about 650°C.

EXAMPLE 3
A Rack of Card-Mounted Components

Figure 3.11 shows a typical rack of card-mounted electronic components. One end of the rack has been removed to show the details of the components on the cards. The components are mounted on one side of 4 in. × 6 in. × 1/16 in. thick circuit board cards. Connections between components are made with printed circuitry on the opposite sides of the card.

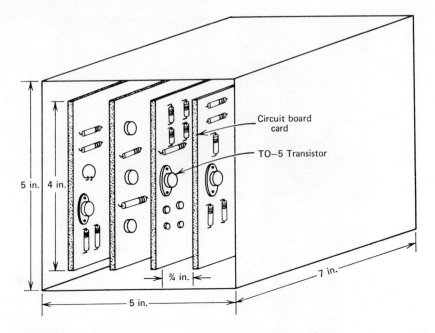

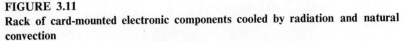

FIGURE 3.11
Rack of card-mounted electronic components cooled by radiation and natural convection

Four cards are mounted in a rack 5 in. wide × 7 in. long × 5 in. high. The top and bottom of the rack are open to allow natural convection cooling of the components. The spacing between the cards is approximately 3/4 in. when allowance is made for the height of the components mounted on the card surface.

Forced air cooling of this same configuration is discussed in Example 3 of Chapter 4. The same configuration will be considered again in Example 2 of Chapter 6, where cooling will be by liquid evaporation in a liquid filled package. By using the same configuration for each of these examples, the relative advantages and disadvantages of each cooling method can be compared.

In this present example, only cooling by natural convection needs to be considered. Radiation can be neglected because of the mutual shielding of the circuit board cards. Considering the cards as an array of fins, the ratio of their height to spacing is 4 in. divided by 3/4 in. or 5.3, and from Figure 3.4, the effective radiated power is only 25 % that of an unshielded radiator. Also, the emissivity of the transistor cases and other components mounted on the cards is low. Therefore, radiation cooling can be neglected.

The power that can be dissipated by a TO-5 transistor mounted in one of the cards will be calculated.

The transistor case temperature will be 70°C, and the temperature of the surrounding air will be 20°C. The temperature difference between the transistor and the surrounding air is therefore 50°C.

The convective air flow pattern will be controlled primarily by the card dimensions themselves. The length over which the air moves is the vertical height of the card of 4 in.

From Equation 3.2:

$$Q = Q_c F_1 F_2 S$$
$$= .3 \text{ W/in}^2 \times .7 \times 1 \times .30 \text{ in}^2$$
$$= .06 \text{ W}.$$

Q_c is the power that would be transferred by natural convection from a 1 in. high vertical surface. The temperature difference between the transistor and the surrounding air is 50°C, and from Figure 3.6, Q_c is .3 W/in^2. F_1 is the correction factor for surface length and orientation. For a 4 in. high vertical surface, $F_1 = .7$ from Figure 3.7. At sea level, $F_2 = 1$. The surface area of the TO-5 transistor is:

$$\text{Top} \quad (\pi/4)\,(.3 \text{ in.})^2 \qquad = .07 \text{ in}^2$$
$$\underline{\text{Side} \quad \pi \times .3 \text{ in.} \times 1/4 \text{ in.} = .23 \text{ in}^2}$$
$$.30 \text{ in}^2$$

Therefore, the power that can be transferred from the transistor with a case temperature of 70°C by natural convection when it is mounted on a circuit board card in the rack is only .06 W. If more power must be transferred by natural convection from this card mounted transistor, its surface area must be increased by the use of a small finned heat sink.

EXAMPLE 4
A Packaged Airborne Electronics Equipment

The package which contains the electronic equipment often serves as an effective radiator for transferring heat by radiation and natural convection.

In this example, the heat that can be transferred from an airborne electronics equipment will be calculated under the following conditions.

Temperature of box surface	50°C
Temperature of surroundings	20°C
Altitude	30,000 ft
Dimensions	5 in. × 5 in. × 8 in. long

The first step is to calculate the heat transferred by radiation, using Equation 3.1 and Figure 3.3. From Figure 3.3B, when the package surface is at 50°C, the radiated power is .14 W/in². The box should be painted to increase its emissivity to .9.

The total surface area is:

2 sides + top + bottom + 2 ends = 2 × 5 in. × 8 in.

$$+ 5 \text{ in.} \times 8 \text{ in.} + 5 \text{ in.} \times 8 \text{ in.} + 2 \times 5 \text{ in.} \times 5 \text{ in.} = 210 \text{ in}^2.$$

Using Equation 3.1:

$$Q = Q \, \varepsilon RS$$
$$= .14 \text{ W/in}^2 \times .9 \times 1 \times 210 \text{ in}^2$$
$$= 26.5 \text{ W.}$$

The next step is to calculate the heat transferred by natural convection using Equation 3.2 and Figures 3.6, 3.7, and 3.8.

The temperature difference between the package surface and the surroundings is 50°C − 20°C = 30°C. From Figure 3.6, the heat transferred by convection is .15 W/in² for a 1 in. vertical surface at sea level.

The effective length along which air moves is 5 in. for the vertical sides and ends of the package. From Figure 3.7, the relative effectiveness factor due to surface dimensions and orientation is .67.

The effective length for the top surface is (5 in. × 8 in.)/(5 in. + 8 in.) = 3 in. and from Figure 3.7, $R_1 = .76$. For the bottom surface, $R_1 = .38$.

The effect of 30,000 ft altitude operation is given by the altitude reduction factor R_2, which is .55 from Figure 3.8 at an altitude of 30,000 ft. The power transferred by natural convection from each surface should now be calculated using Equation 3.2.

For the top surface:

$$Q = Q_c F_1 F_2 S$$
$$= .15 \text{ W/in}^2 \times .76 \times .55 \times 5 \text{ in.} \times 8 \text{ in.}$$
$$= 2.5 \text{ W.}$$

Calculations made in the same manner for the other surfaces give the following results.

Surface	Q_c (W/in²)	F_1	F_2	Area (in²)	Power (W)
Top	.15	.76	.55	5 × 8	2.5
Bottom	.15	.38	.55	5 × 8	1.3
2 Sides	.15	.67	.55	2(5 × 8)	4.4
2 Ends	.15	.67	.55	2(5 × 5)	2.8
				TOTAL	11.0

The total heat transferred from the package at 50°C is therefore:

Radiation   26.5 W
Convection   11.0 W
37.5 W

The components inside the package must, of course, be hotter than the package surface to be able to transfer their heat to the package. The heat may be transferred from the components to the package walls by conduction, radiation and natural convection, or any of the heat transfer means. In Example 2 of Chapter 6, the transfer of heat from the components to the package walls by evaporation will be discussed.

3.5 REFERENCES

Useful references for cooling of electronic equipment by radiation and natural convection are as follows:

1. Krauss, A. D., *Cooling Electronic Equipment*, Prentice Hall, Inc., Englewood Cliffs, N.J., 1954, Chapter 2, pp. 43–51, Chapter 3, pp. 52–69.

2. *Guide Manual of Cooling Methods for Electronic Equipment*, NAVSHIPS 900,190, U.S. Government Printing Office, Washington, D.C., pp. 28–39.

3. McAdams, W. H., *Heat Transmission*, McGraw-Hill, New York, 1954, Chapter 4, pp. 55–82, Chapter 7, pp. 165–183.

4. Goldman, W., *A Primer on Heat Sinking*, Wakefield Engineering, Inc., Wakefield, Mass., 1966.

5. *Heat Sinks*, Thermalloy, Inc., Dallas, Texas, 1972.

Reference 1 provides the basic equations from which the formulas and graphs of this Chapter are derived. Chapter 2 covers natural convection. Chapter 3 covers radiation.

Reference 2 presents the same information as Reference 1, and, in addition, presents some particularly useful nomographs for calculation.

Reference 3 provides the same basic equations of References 1 and 2.

Reference 4 provides practical rules for the use of radiation and natural convection cooled heat sinks.

Reference 5 provides application information and a detailed listing of representative heat sinks for electronic equipment.

4

Forced Air Cooling

An order of magnitude increase in heat transfer can be achieved by blowing air over the electronic component, rather than relying on radiation and natural convection. The price that must be paid for this increased cooling is:

- Increased system complexity, because a fan and its associated equipment (such as ducting, dust filters, and interlocks) are required to force the air over the component.
- Reduced electrical efficiency for the system, because the fan requires electrical power.
- Increased vibration and acoustical noise.

Obviously, heat transfer by radiation and natural convection should be used wherever possible, to avoid the increased system complexity of forced air cooling. However, when the heat that must be transferred is greater than 50 W, the size of the fins required for radiation and natural convection becomes excessive, and forced air cooling with all of its complexity often provides the best system design.

Although more complicated than radiation and natural convection, forced air cooling is certainly simpler than forced liquid cooling, because a supply of cooling air is readily available and air does not have the freezing, boiling, or dripping problems of liquids.

The design of forced air cooling poses two problems:

1. Design of the cooling fins on the electronic component.
2. Choice of the fan or blower.

These two problems must be solved jointly. The amount of air flow that a particular fan can provide is determined by the pressure into which the fan must work. Both the amount of heat transfer that can be obtained from forced air cooling and the pressure required to force air through the cooling

64

fins depends on air flow and fin geometry. Consequently, the fin design must be made in conjunction with the choice of fan. This tradeoff is illustrated in Section 4.1, where a typical forced air cooling design is described.

Design information for calculating the temperature rise of the component with forced air cooling and the pressure drop required to force the air through the fins is presented in Sections 4.2 through 4.5. The temperature rise of the cooling air itself, which is independent of fin size or geometry, is discussed in Section 4.2. Information for designing air cooling fins is given in Section 4.3. Cooling of electronic components in racks and cabinets is described in Section 4.4.

The effect of altitude and air temperatures different from the standard sea level and room temperature environment is considered in Section 4.5.

Fans for the forced air cooling of electronic equipment are discussed in Section 4.6. This discussion includes:

1. The pressure versus airflow characteristics of the fans.
2. The size, electrical power, acoustic noise, life, and cost of the fans.
3. Auxiliary equipment such as dust filters, ducting, and interlocks.

Examples of forced air cooling designs of electronic equipment are presented in Section 4.7 to illustrate the use of the equations, figures and fan specifications of this chapter.

4.1   A TYPICAL FORCED AIR COOLING DESIGN

Figure 4.1 shows a typical forced air cooled heat sink. The conduction of heat from the electronic component to the fins has been discussed in detail in Chapter 2. A fan blows air through the fins, and the heat is transferred from the fins to the air stream. The temperature of the air rises as it absorbs heat, and the fin temperature rises to be hotter than the air in order to transfer heat.

The temperature rise of the fins above the inlet air temperature is shown in Figure 4.2 for the particular heat sink shown in Figure 4.1. This curve shows the measured temperature rise at the base of the fins per watt of power transferred as a function of air flow rate in cubic feet per minute (CFM).

At a flow rate of 18 CFM, Figure 4.2 shows that the temperature rise of the fins above the inlet air temperature is .5°C/W. If the inlet air from the surroundings is at 20°C and the fins are at 100°C, the temperature difference is 80°C and 160 W can be transferred.

If the air flow through the fins is increased to 45 CFM, the heat transfer rate doubles. With an 80°C temperature difference between the fins and the inlet air, 320 W can be transferred. This is significantly more heat than can be transferred by radiation and natural convection from a heat sink of the same size.

FIGURE 4.1
Forced air cooled heat sink (Photo courtesy of Wakefield
Engineering, Inc.)

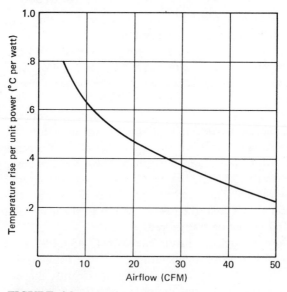

FIGURE 4.2
Temperature rise of forced air cooled heat sink as a
function of airflow

For example, the heat sink discussed in Section 3.1 of Chapter 3 and shown in Figure 3.1, which is designed for cooling by radiation and natural convection, occupies three times the volume of the forced air cooled heat sink and fan combination. At the same temperature difference of 80°C, it can transfer only 60 W, or less than one-fifth the power of the forced air cooled sink.

The foregoing discussion suggests that as great an air flow as possible should be used to maximize the heat transfer capability of the fins. However, the greater the air flow, the greater the pressure required to force the air through the fins. Pressure as a function of air flow is shown in Figure 4.3

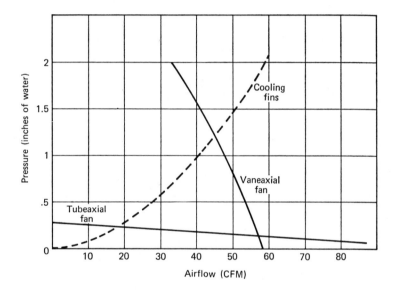

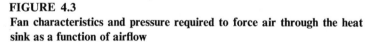

FIGURE 4.3
Fan characteristics and pressure required to force air through the heat sink as a function of airflow

for the particular heat sink shown in Figure 4.1. Pressure is expressed in "inches of water," which is the height of a column of water that the pressure will support. This choice of unit is used because fan pressure capability is commonly expressed in this unit, and because pressure is easily measured with a water column manometer (as described in Chapter 10).

As Figure 4.3 shows, at an air flow of 18 CFM, the pressure required to force the air through the cooling fins is only .2 in. of water. The required pressure increases rapidly with flow rate. At 45 CFM, the required pressure is 1.2 in. of water.

Pressure versus air flow characteristics of two fans which are commonly used in electronic equipment are superimposed over the fin characteristics of Figure 4.3. These particular fans are described in detail in Section 4.6, and are illustrated in Figures 4.13 and 4.14. The intersection of the pressure versus flow characteristics of the cooling fins with the pressure versus flow characteristics of the fans determines the actual air flow through the fins and thereby determines their heat transfer capability.

The higher pressure capabilities of the vaneaxial fan allow it to force 45 CFM of air through the cooling fins, and with this air flow the fin assembly has a heat transfer capability of .25°C/W.

In contrast, the tubeaxial fan has a much lower pressure capability and can force only 18 CFM through the cooling fins. Therefore, when the same fin assembly is used with the tubeaxial fan, its heat transfer capability is only one-half as good as when it is used with the vaneaxial fan.

From a consideration of only pressure versus air flow capabilities, the higher pressure vaneaxial fan would be the best choice. However, the choice of cooling fan is dictated by other requirements besides pressure versus air flow characteristics, namely:

1. Size and weight.
2. Power consumption.
3. Acoustical noise.
4. Life.
5. Cost.

As will be discussed in detail in Section 4.6, the vaneaxial fan has a better pressure versus air flow characteristic and is smaller in size and weight, but the tubeaxial fan offers the advantages of less power consumption, lower acoustical noise, longer life, and lower cost.

The effectiveness of the various methods of heat transfer were compared in Figure 1.2 of Chapter 1. The value shown for forced air cooling was taken from the cooling fin design shown in Figure 4.1. At an air flow of 45 CFM, which can be supplied by the vaneaxial fan, the temperature rise of the fins is .25°C/W. The total surface area of the fins is 13 ducts × 2 sides/duct × 1.5 in. × 1 in. = 40 in^2. If the fin temperature is 100°C and the inlet air temperature is 20°C, the heat transferred is 8 W/in^2.

Air cooling fins can be spaced much closer together than fins for radiation and natural convection cooling because there is no shielding problem. Consequently, forced air cooling designs can achieve much more fin surface in a given volume as well as greater heat transfer per unit fin area than radiation and natural convection designs.

4.2   TEMPERATURE RISE OF THE COOLING AIR

The temperature rise of the cooling fins above the inlet air temperature consists of two parts, as shown by the following formula:

$$T \text{ (fin)} = \Delta T \text{ (air)} + \Delta T \text{ (fin-air)} + T \text{ (surroundings)} \qquad (4.1)$$

where:

T (fin)	is the temperature of the cooling fin surface (°C).
ΔT (air)	is the temperature rise of the air as it absorbs heat from the cooling fins (°C).
ΔT (fin-air)	is the temperature rise of the cooling fin above the air in the cooling duct (°C).
T (surround-ings)	is the temperature of the inlet air, which is equal to the temperature of the surroundings (°C).

The air temperature increases as the air passes along the fins because the air is absorbing power. This is the term ΔT (air) of Equation 4.1. The fin must be hotter than the cooling air to transfer heat. This is the term ΔT (fin-air).

The temperature rise of the air as it absorbs heat from the cooling fins is independent of the geometry or dimensions of the fins. The air temperature rise depends only on the power that is transferred and the total air flow rate, and is given by the following formula:

$$\frac{\Delta T \text{ (air)}}{Q} = \frac{1.73}{f} \qquad (4.2)$$

where:

ΔT (air)	is the temperature rise of the air in absorbing heat from the cooling fins (°C).
Q	is the total power that is being transferred (watts).
f	is the total air flow through the fins (CFM).

Equation 4.2 applies only to air at sea level at a temperature of 20°C. Modifications of the constant of the equation to include other altitudes and air temperatures are discussed in detail in Section 4.5.

Equation 4.2 is illustrated graphically in Figure 4.4, where the temperature rise of the cooling air is shown as a function of the power that is being transferred with air flow as a parameter. Note that with an air flow of 50 CFM, 500 W of power can be transferred with an air temperature rise of only 17°C. As will be discussed in Section 4.6, fans are available which can provide air flows as high as 500 CFM. With this air flow rate, 10 kW of heat can be transferred with an air temperature rise of only 35°C.

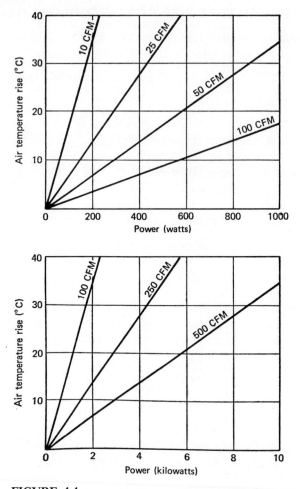

FIGURE 4.4
Temperature rise of cooling air as a function of power
absorbed

4.3 DESIGN EQUATIONS FOR AIR COOLING FINS

The design of forced air cooling of electronic equipment has two parts.
The cooling fin dimensions must be selected to obtain sufficient cooling and
at the same time keep the pressure drop through the fins low enough so that
the required air flow can be supplied by a suitable fan. The fan must be
chosen not only for its pressure versus air flow characteristics but also to meet
other system requirements of size, weight, power consumption, acoustical
noise, life and cost.

In this section, design formulas are presented for estimating the heat transfer capability and the pressure requirements as a function of fin geometry and air flow under room temperature and sea level operating conditions. Modifications to the formulas to account for the effect of varying air density at other temperatures and altitudes are presented in Section 4.5. The problem of fan selection is discussed in Section 4.6.

As shown in Equation 4.1, the temperature rise of the cooling fins above the inlet air temperature consists of two parts—the temperature rise of the air as it absorbs power and the temperature rise of the fins above the air. The temperature rise of the air, as discussed in the previous section, is independent of fin geometry. In any fin design, the fins will get hot enough to transfer the heat being generated in the electronic components to the air stream. However, unless the fins are designed properly, this temperature rise may be excessive.

Fin Temperature Rise

The temperature rise required to transfer a given amount of heat from the fins depends upon the velocity of the air moving past the fins, and the air velocity in turn depends on the fin geometry and dimensions and on the air flow rate. The temperature rise is related to these factors by the following formula:

$$\frac{\Delta T \text{ (fin-air)}}{Q} = \frac{140\,w}{n^{.2} z^{.2} f^{.8} L} \tag{4.3}$$

where:

ΔT (fin-air) is the temperature rise of the fin surface above the air flowing through the cooling ducts between the fins (°C).
Q is the total power that is being transferred (watts).
w is the width of the ducts, i.e. the spacing between fins (inch).
z is the height of the fins above the base (inch).
L is the length of the fins along the direction of air flow (inch).
n is the number of ducts through which the air flows.
f is the total air flow through all the ducts (CFM).

The geometry of the cooling fins is shown in Figure 4.5, and this figure defines the critical fin dimensions w, z, and L of Equation 4.3.

Forced air flow is a complicated hydrodynamic problem which is affected by many factors. Several simplifying assumptions have been made in deriving Equation 4.3, so it gives only an approximate result. In spite of its approximate nature, the equation is useful in estimating cooling capability and in

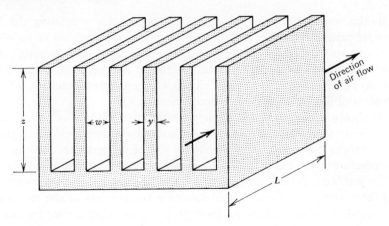

FIGURE 4.5
Geometry of forced air cooled heat sink and important dimensions

deciding how changes in fin dimensions will affect cooling capability. The degree of correlation between Equation 4.3 and measured results is shown in the examples of Section 4.7. Ultimately a cooling design obtained by the use of Equation 4.3 must be built and experimentally evaluated. Suitable measurement techniques for forced air cooling designs are described in detail in Chapter 10.

The major assumptions used in deriving Equation 4.3 are as follows:

1. The average air temperature is 20°C and the altitude is at sea level.
2. The air flow is turbulent.
3. The ducts are narrow compared to their height above the base.

The first assumption of sea level operation and 20°C air affects only the coefficient of the equation. Corrections to this coefficient for other altitudes and air temperatures are presented in Section 4.5.

Is is useful to examine Equation 4.3 to determine the major factors affecting heat transfer. The temperature rise of the fins above the air is directly proportional to the duct width, i.e., the spacing between the fins, so reducing this dimension will correspondingly reduce the temperature difference between the fin surface and the air. In contrast, the temperature rise of the fin is almost independent of duct height.

The temperature rise of the fin is inversely proportional to the length of the fin in the direction of air flow, so, for example, doubling the length of the fin would cut the temperature rise in half.

The fin temperature rise is inversely proportional to the .8 power of air flow. Consequently, to reduce fin temperature rise by a factor of two, the air flow rate must be more than doubled. This fact is evident from the experimental results shown in Figure 4.2, where air flow must be increased from 18 CFM to 45 CFM, i.e., 2.5 times, to reduce the temperature rise per watt by a factor of two. The number of ducts that the air flow is divided into has very little effect on heat transfer capability as long as the duct width (i.e., the fin spacing) is constant, since heat transfer capability depends on the number of ducts to the .2 power. The total air flow could all be forced through a single duct or it could be divided up into many ducts, with only a small change in fin temperature rise.

Equation 4.3 assumes a perfectly conducting fin. The fin temperature rise calculated from Equation 4.3 must be corrected to account for the finite thermal conductivity of the fin, as outlined in Section 2.6 of Chapter 2.

Pressure

Unfortunately, the fin geometry and dimensions cannot be selected arbitrarily to minimize temperature rise, because they also affect the pressure required to force air through the fins. The minimizing of the air pressure drop is an important step in reducing fin temperature. As shown by the formulas of the previous paragraphs, the temperature rise of the fins depends on the amount of air flow. However, the amount of air flow that can be forced through the fins by a particular fan depends on the pressure versus air flow characteristics of the design. Therefore, reducing the air pressure drop' through the equipment is extremely important.

The total pressure into which the fan must work is determined not only by the cooling fins themselves but by all of the following elements:

- Entry port.
- Dust filter.
- Inlet ducting.
- Cooling fins.
- Outlet ducting.
- Exit port.

Of all of these elements that are in the air flow path, only the cooling fins contribute to cooling, but all contribute to the pressure drop.

An exact calculation of the pressure drops through each element of the system is extremely difficult. However, the application of three general rules allow unnecessary pressure drops to be minimized. These rules are:

1. The pressure drop through any structure is inversely proportional to the square of its cross-sectional area. For example, if the cross-sectional area of a duct is doubled, the pressure drop through it is reduced to one-fourth.

2. The pressure drop through any element is approximately directly proportional to the air flow squared. This fact is evident from the pressure versus air flow characteristics shown in Figure 4.3. At 18 CFM the pressure drop is approximately .2 in. of water. When the air flow is doubled to 36 CFM, the pressure drop increases to .8 in. of water, a four times increase.

3. The pressure drop through any structure is proportional to the length of the structure (except that for short structures the fixed entry and exit pressure drops may be much larger than the pressure drop due to air flow through the structure).

Applying the first of the above rules, the cross-sectional area of the entry port, inlet ducting, outlet ducting, and the exit port should be made large enough and without sharp bends or abrupt transitions so that the pressure drop through all of these elements (which do not contribute to cooling) can be negligible. This first step must be taken in any forced air cooling design.

In a well designed equipment, the major part of the pressure drop will be the cooling fins themselves. The geometry and dimensions of the fins affect both the heat transfer and the pressure drop, and a compromise design between the two must often be used. An approximate formula for estimating the pressure drop through cooling fins as a function of their dimensions is as follows:

$$\Delta p = \frac{\left(\dfrac{f}{n}\right)^2}{(wz)^2}\left[1 + .01\,\frac{L}{w}\right] \times 10^{-3} \tag{4.4}$$

where:

Δp is the pressure drop through the fins (inches of water).
All other terms are as defined for Equation 4.3.

Equation 4.4 is based on the same assumptions that were used for Equation 4.3, that is, sea level operation with 20°C inlet air and turbulent flow. A multiplying constant which permits Equation 4.4 to be used at other altitudes and inlet air temperatures is discussed in Section 4.5.

As would be expected, Equation 4.4 illustrates the three general rules for air pressure drop stated above. The pressure drop is inversely proportional to the square of the cross-sectional area of the ducts (which is wz) and directly proportional to the square of the air flow rate per duct. The cross-sectional area of the ducts cannot be made arbitrarily large to reduce pressure drop because duct dimensions affect both heat transfer as well as pressure. Note, however, by comparing Equations 4.3 and 4.4, that the fin temperature depends directly on duct width but only slightly on duct height, whereas pressure drop is equally affected by both. Therefore, fin height can be increased to reduce the pressure drop without significantly degrading the heat transfer capability of the fins.

Note also that although the fin temperature rise is fairly insensitive to the number of fins used, the pressure drop is very sensitive, varying inversely as the square of the number of fins.

The first term in the brackets accounts for the transition into and out of the fin structure. The second term accounts for the pressure drop through the fins themselves. Note that until the length of the fins is 100 times their spacing, the pressure drop due to the transitions into and out of the finned structure is greater than the pressure drop due to the air flowing through the fins.

Certain additional simplifying assumptions have been used for the terms in the bracket of Equation 4.4. The first term is actually a function of the ratio of fin thickness to duct width, but a fixed value for the case when the thickness to width ratio is one-third has been used. The second term depends slightly on flow rate and this variation has been ignored. The principal use of Equation 4.4 should be in obtaining a rough estimate of pressure drop through the fins and in determining how changes in fin dimensions will affect the pressure drop. In any actual equipment, the pressure drop must be measured using the techniques described in Chapter 10.

4.4   FORCED AIR COOLING OF RACKS AND CABINETS

A common means of cooling electronic equipment is by blowing air through a rack or cabinet to cool all of the electronic components mounted in the cabinet. Figure 4.6 shows a typical electronic equipment cabinet with a blower mounted in the bottom. The air flow is shown schematically.

The air temperature increases as it absorbs power from the components, and therefore the air is the hottest at the top of the cabinet. The temperature rise of the exit air is independent of component shapes, sizes or locations and depends only on the total power absorbed from the components and the total air flow rate. This air temperature rise can be calculated from Equation 4.2, which was shown graphically in Figure 4.4.

The electronic components must, of course, be hotter than the cooling air in order to transfer heat to the air. Because the air temperature increases as it circulates through the cabinet, the components near the top of the cabinet will be hotter than the components near the bottom of the cabinet even if their power dissipation is the same. Enough air should therefore be blown through the cabinet to limit the air temperature rise to about 25°C. From Equation 4.2, if a total power of 1000 W is dissipated in the cabinet, the air flow should be 70 CFM.

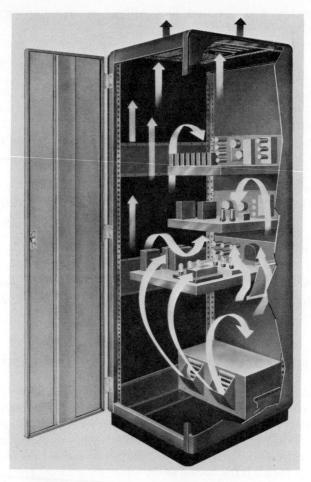

FIGURE 4.6
Airflow in a forced air cooled rack of electronics
equipment (Photo courtesy of McLean Engineering
Laboratories)

The amount of air that can be blown through the cabinet depends on the
pressure versus air flow characteristics of the cabinet and the pressure versus
air flow capabilities of the fan. A simple formula is not available for calculating the air pressure required to flush a given amount of air through an
electronics cabinet because of the complicated nature of the air flow. However, an estimate can be made from the empirical data presented in Figure
4.7, which shows pressure versus air flow characteristics for five typical
electronics racks and cabinets.

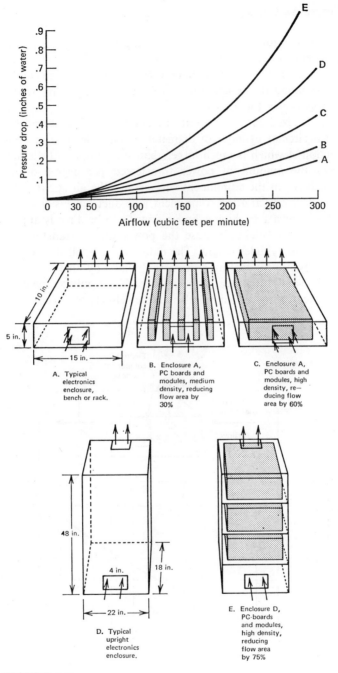

FIGURE 4.7
Pressure drop as a function of airflow for typical electronic enclosures (Courtesy of Pamotor, Inc.)

Likewise, a simple calculation of the temperature rise of each component as it transfers its heat to the air stream is not possible because the air flow is so complex. However, an estimate of the temperature rise can be obtained by approximating the actual situation for each component by the idealized geometry shown in Figure 4.8.

In this idealized case, the electronic component is assumed to be in the shape of a circular cylinder of diameter d and is located in a duct with a cross section A which is much larger than the component diameter. Figure 4.8 shows the amount of power transferred per degree of temperature difference between the component surface and the air stream as a function of the air flow rate and the component and duct dimensions. The curve of Figure 4.8 is for a 1 in. high cylinder, but it can be directly applied to components of other heights because the power that is transferred is directly proportional to cylinder height. The data can be applied to components of any shape by approximating their actual shape by an appropriate circular cylinder. The amount of heat transfer depends not only on the dimensions of the component being cooled but also on the velocity of the cooling air, which

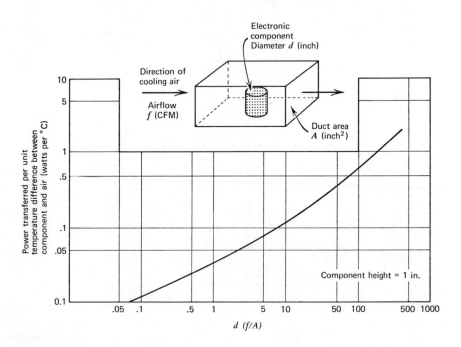

FIGURE 4.8
Power transferred from a cylindrical component in an airstream

is determined by the volume of air flow and the cross-sectional area of the duct in the cabinet through which the air is blown.

The use of Figure 4.8 is illustrated by the following example. Assume the component to be cooled is an electron tube with a diameter of 3/4 in. and a height of 2 in. An air flow of 25 CFM is blown through a 5 in. × 5 in. duct in which the tube is located. $d(f/A) = 3/4$ in. × 25 CFM/5 in. × 5 in. = .75 and from Figure 4.8:

$$\frac{Q}{\Delta T} \text{ is .03 W/}^\circ\text{C.}$$

Since the cylinder is 2 in. high and Figure 4.8 is for a 1 in. high cylinder, twice as much power will be transferred, that is, .06 W/°C. If the air temperature is 20°C and the tube surface is at 100°C, the temperature difference is 80°C, and approximately 5 W of power will be transferred.

This type of air cooling is less effective than if the component were mounted on a finned heat sink such as was shown in Figure 4.1, but it is more effective than if only natural convection were used.

For comparison, the amount of heat transferred from the tube by natural convection can be calculated from the design information of Chapter 3. When the tube surface is at 100°C and the surroundings are at 20°C, .4 W/in² will be transferred by natural convection from a 2 in. vertical surface. The area of the tube surface is 5 in², so the total power transferred by natural convection is 2 W or only 40% as much as when air is flushed past the tube. Note, however, that if the air flow were less, or if the tube were located in a larger duct with the same volume of air flow, the cooling that could be obtained would be less.

In most practical cases, the cooling that can be obtained for any individual component by blowing air through a rack or cabinet is only slightly better than can be obtained by natural convection. Any attempt to duct the air around the components will raise the air pressure requirements so that a reasonable fan cannot blow enough total air through the cabinet. A good general rule therefore is to design each chassis so that all components on it are satisfactorily cooled by conduction, radiation, and natural convection, and depend on the fan only to bring sufficient air into the cabinet. In other words, the individual chassis should be designed to operate on the bench without forced air cooling. The fan is then used to insure that enough air is provided so that several chassis can operate together inside the cabinet.

Air should always be blown into the cabinet, as shown in Figure 4.6, rather than sucked in, to keep dust out of the cabinet. A dust filter should be mounted at the fan inlet.

4.5   THE EFFECT OF ALTITUDE AND INLET AIR TEMPERATURE

The design equations and curves given in the previous sections were all for the case of sea level operation with an inlet air temperature of 20°C. At other altitudes and air temperatures, the amount of cooling and the pressure drop obtained with a given volume of air flow will be different, because the air density and the thermal properties of air are different.

For Equations 4.2 through 4.4, which give the temperature rise of the cooling air as it absorbs power, the temperature rise of the cooling fins above

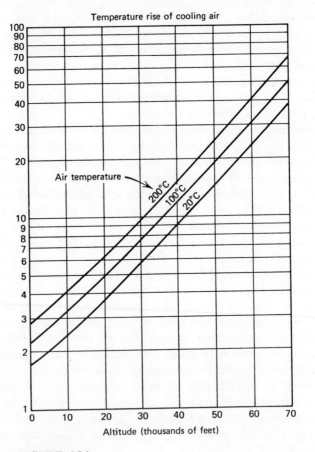

FIGURE 4.9A
Coefficient of Equation 4.2 as a function of altitude and air temperature

the air, and the pressure drop through the fins, respectively, the value of air density and the other air thermal properties affect only the constant multiplying factor of the equations. Consequently, Equations 4.2 through 4.4 can be used at any altitude or air inlet temperature simply by using a different constant. Appropriate values of these constants as a function of altitude, with inlet air temperature as a parameter, are shown in Figure 4.9. The value of these constants at sea level and an inlet air temperature of 20°C from Figure 4.9 is, of course, the same as given with the equations in the previous section.

As altitude and inlet air temperature increase, the temperature rise of both the cooling air and the fins increases. For example, when altitude increases to 35,000 ft, the temperature rise of a given volume flow of cooling air increases four times over the temperature rise that occurs at sea level. The temperature rise of the fins above the cooling air increases three times.

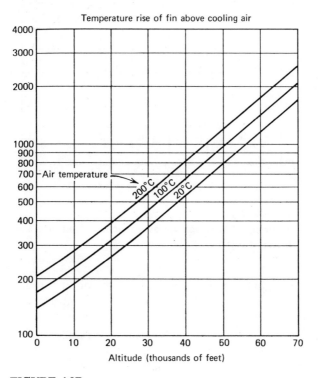

FIGURE 4.9B
Coefficient of Equation 4.3 as a function of altitude and air temperature

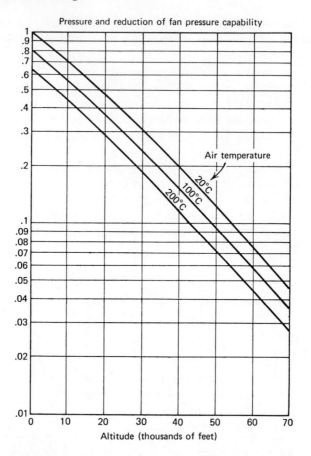

FIGURE 4.9C
Coefficient of Equation 4.4 as a function of altitude and air temperature

The pressure required to force a given volume of air through the cooling fins decreases with increasing altitude and inlet air temperature, as shown by the curves of Figure 4.9. However, the pressure versus air flow capability of most fans decreases by the same ratio with increasing altitude and inlet air temperature, with the result that the air flow that a particular fan can blow through a particular fin assembly is approximately constant at any altitude or inlet air temperature.

The effect of altitude on the performance of the forced air cooled heat sink described in Section 4.1 is shown in Figure 4.10. The temperature rise of the fins and the pressure required to force air through them is shown as a function

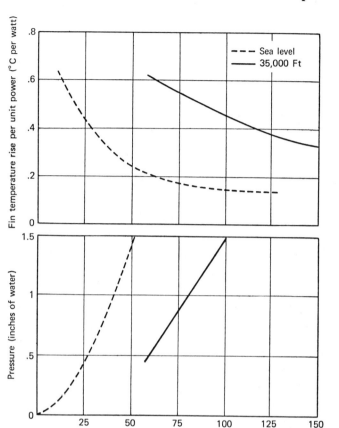

FIGURE 4.10
A comparison of cooling fin temperature and pressure at sea level
and at an altitude of 35,000 ft

of air flow at both sea level and at an altitude of 35,000 ft. (The sea level
curves are the same as were shown previously in Figures 4.2 and 4.3.) As
expected, the fin temperature rise is greater at 35,000 ft than at sea level for a
given air flow and the pressure required to force air through the fins is less.

Estimating High Altitude Performance from Sea Level Measurements

The equations presented in Sections 4.2 and 4.3 for temperature rise and
pressure give only approximate answers. Ultimately a forced air cooling
design obtained by the use of these equations must be built and experimentally measured. The experimental measurements at sea level and with

room temperature inlet air are easy to make, and are described in detail in Chapter 10.

These same measurements are much more difficult to make for high altitude conditions and for inlet air temperatures different from room temperature. For these cases a special temperature-altitude chamber must be used.

Fortunately, this experimental complication can be avoided by measuring the performance of the forced air cooling design at sea level with room temperature air and then extrapolating the measured results to the desired altitude and inlet air temperature using the correction factors of Figure 4.9. The correction factors involve only the density and thermal properties of air and these are well established as a function of altitude and air temperature. The effect of the air flow pattern on cooling and pressure drop, which can only be approximately calculated from the geometry and dimensions of the fins, can be exactly determined from the sea level, room temperature measurements.

The performance of the forced air cooled heat sink described in Section 4.1 at an altitude of 35,000 ft, which is shown in Figure 4.10, was deduced in just this way. For example, the cooling fin temperature rise at 60 CFM at sea level was measured to be .2°C/W. From Figure 4.9, operation at 35,000 ft would increase the fin temperature by $450/140 = 3$ times, so the fin temperature would be .6°/W. Similarly, the measured sea level pressure at 60 CFM is 2 in. of water. From Figure 4.9, this pressure would be reduced to .25 × 2 in. of water or .5 in. of water at an altitude of 35,000 ft.

One caution should be exercised in using Figure 4.9 to extrapolate measured results to other altitudes and inlet air temperatures. The extrapolation assumes that the air flow remains turbulent. Greater air flow is required for turbulent flow at 35,000 ft than at sea level. For this reason, the extrapolation for 35,000 ft operation has not been extended to low air flow rates in Figure 4.10.

Fan Selection for High Altitude Operation

The pressure that a particular fan can supply at a given air flow is proportional to the air density and so decreases with increasing altitude and air inlet temperature as shown in Figure 4.9.

Fan pressure versus air flow characteristics are normally supplied by fan manufacturers for sea level and room temperature operating conditions. The performance of a fan at any altitude and air inlet temperature may be extrapolated from the manufacturers data by multiplying the pressure at each flow rate by the fan pressure reduction factor shown in Figure 4.9. The use of this approach is illustrated in Figure 4.11, which shows the pressure versus air flow characteristics of the finned heat sink described in Section 4.1 and the pressure versus air flow characteristics of a 3 in. diameter vaneaxial fan at an altitude of 35,000 ft and at sea level. The pressure versus air flow characteristics of the finned heat sink at 35,000 ft were shown previously in Figure 4.10 and were deduced from sea level measurements as described in the previous paragraph. The problem is to select an appropriate fan for use with these fins at 35,000 ft. The dotted curve shows the sea level characteristics of the 3 in. diameter vaneaxial fan and is taken from the fan manufacturers published data. The performance of this fan at 35,000 ft is shown by the solid curve and is obtained by multiplying its sea level pressure capability at each flow rate by the appropriate reduction factor from Figure 4.9. As Figure 4.11 shows, at 35,000 ft the pressure versus air flow characteristics of the fan and the cooling fins intersect at an air flow of 70 CFM. This is the air flow at which the combination will operate at 35,000 ft, and from Figure 4.10 the temperature rise of the fins will be .57°C/W.

Airborne electronics equipment must operate over a range of altitudes from sea level to the maximum altitude capability of the aircraft. The most difficult cooling problem usually occurs at the maximum altitude, and so the cooling fins must be designed and the fan selected for this altitude as described in this section. At lower altitudes the forced air cooling system will be overdesigned, and the fan will consume an excessive amount of power, because the power requirements of a fan are approximately proportional to

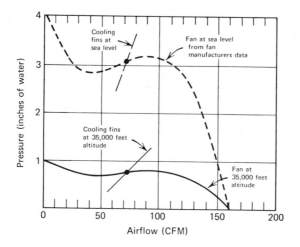

FIGURE 4.11
**A comparison of fan characteristics at sea level and at
an altitude of 35,000 ft**

its pressure-air flow rate product. For these variable altitude applications, special fans are available which reduce their rotation speeds and thereby their air flow rate at low altitude, where sufficient cooling is obtained with low air flows. In this way the power requirements of the fan are kept reasonably constant with altitude.

4.6   FANS FOR COOLING ELECTRONIC EQUIPMENT

The selection of the proper fan for a particular electronic cooling application is as critical as the design of the cooling fins themselves. The amount of cooling that can be obtained depends on the amount of air that can be forced through the fins, and the amount of air is determined by the pressure versus airflow characteristics of both the cooling fins and the fan or blower. Therefore, the choice of fan and the cooling fin design must be done jointly.

The four types of fans that are commonly used for cooling electronic equipment are illustrated in Figures 4.12 through 4.15.

Figure 4.12 shows a "propeller fan." This type of fan provides high air flow volume at low pressures and is best suited for flushing operations in which a large volume of air is blown through a rack or cabinet to directly cool all of the components. The free delivery (i.e., the maximum air flow at zero pressure) that can be obtained from propeller fans depends on the propeller diameter and varies from about 25 CFM for a 3 1/2 in. diameter propeller up to about 1000 CFM for a 10 in. diameter propeller. The particular model shown in Figure 4.12 has a 6 1/2 in. diameter propeller and as shown provides a free delivery of 160 CFM. The maximum pressure that can be developed by these fans with any propeller diameter is only about .5 in. of water.

Figure 4.13 shows a "tubeaxial fan." Its motor is mounted in the fan hub. As a consequence, a tubeaxial fan can develop better pressure and occupies much less axial length than a propeller fan of the same diameter. The fan illustrated in Figure 4.13 is widely used for cooling commercial electronic equipment. It weighs only 1 lb, is quiet, has an expected operating life without servicing of 10 yr, and is low in cost. To make use of all these advantages, the cooling fins that are used with the fan must be carefully designed because the fan has a pressure capability of only about .1 in. of water at high flow rates. A design of cooling fins to optimally use this particular tubeaxial fan is given in Example 2 of Section 4.7.

Figure 4.14 shows a "vaneaxial fan" and pressure versus air flow curves for a 2 in. diameter model and for a 3 in. diameter model. These vaneaxial fans operate at high rotation speeds and consequently have high pressure capabilities and very small sizes.

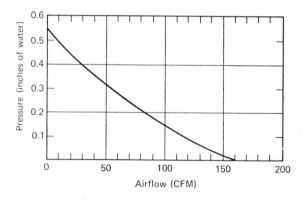

FIGURE 4.12
Propeller fan and its pressure—airflow characteristics
(Photo courtesy of Rotron, Inc.)

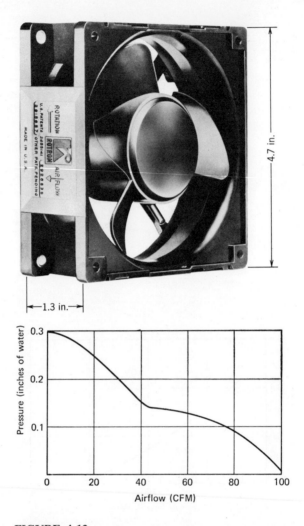

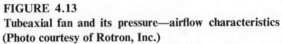

FIGURE 4.13
Tubeaxial fan and its pressure—airflow characteristics
(Photo courtesy of Rotron, Inc.)

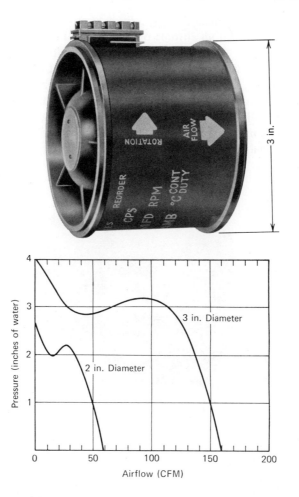

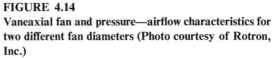

FIGURE 4.14
Vaneaxial fan and pressure—airflow characteristics for two different fan diameters (Photo courtesy of Rotron, Inc.)

At an air flow of 100 CFM, the 3 in. diameter vaneaxial fan has more than ten times the pressure capability of the propeller fan of Figure 4.12 and yet occupies less than one-tenth the volume. This high performance capability is obtained at the expense of high acoustic noise, short life, and the need for 400 cycle power.

The air cooling fins described in Section 4.1 were designed to use the 2 in. diameter vaneaxial fan at sea level and the 3 in. diameter fan at an altitude of 35,000 ft. The performance of the 3 in. diameter vaneaxial fan was compared at sea level and at an altitude of 35,000 ft in Figure 4.11.

Figure 4.15 shows a "squirrel cage centrifugal blower." This blower provides high air flow and high pressure capabilities at low rotation speeds. Squirrel cage blowers with wheel diameters from 1.5 in. to 7.5 in. provide free deliveries of 10 to 700 CFM and operating pressures of 1 to 2 in. of water. The particular squirrel cage blower shown in Figure 4.15 has a 5 in. diameter wheel, and as shown, provides 160 CFM at free delivery. A design of cooling fins on a high power tube to match this blower is given in Example 4 of Section 4.7.

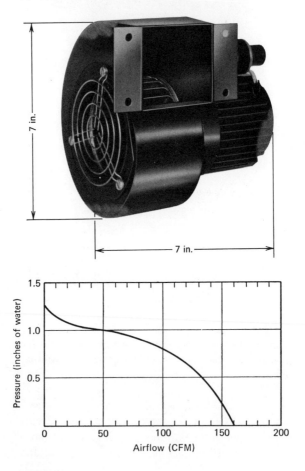

FIGURE 4.15
Squirrel cage blower and its pressure—airflow characteristics (Photo courtesy of Rotron, Inc.)

Selection of the Most Appropriate Fan for a Particular Electronic Cooling Application

The following factors must be considered in selecting the most appropriate fan for a particular electronic cooling application:

1. Pressure versus air flow characteristics.
2. Size and weight.
3. Power requirements.
4. Acoustic noise.
5. Life.
6. Cost.

To give some insight into the tradeoffs between these factors, the pressure versus airflow characteristics of the four fans shown in Figures 4.12 through 4.15 are compared in Figure 4.16, and the other important factors for fan selection are tabulated in Table 4.1.

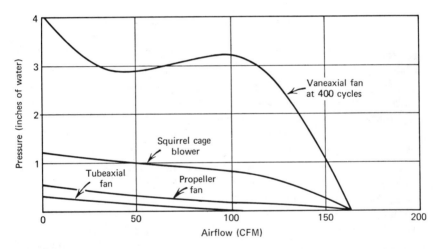

FIGURE 4.16
Comparison of the pressure—airflow characteristics of propeller fan, tubeaxial fan, vaneaxial fan, and squirrel cage blower

Many different levels of performance can be obtained from a particular type of fan by using different wheel diameters, different rotation speeds, different motors, bearings, etc. Therefore, Table 4.1 does not really provide a comparison between types of fans, but only a comparison between the four particular fans illustrated in Figures 4.12 through 4.15.

The propeller fan, the squirrel cage blower, and the vaneaxial fan all provide about the same free delivery of 160 CFM. The squirrel cage blower

TABLE 4.1
Comparison of Fan Types

Fan Type	Propeller Fan	Tubeaxial Fan	Vaneaxial Fan	Squirrel Cage Blower
Specific Model	Rotron MF 5501	Rotron Muffin	Rotron Aximax 3	Rotron VS38A2
Size	8 in. Dia. × 5 in.	4.7 in. × 4.7 in. ×	3 in. Dia. × 2.5 in.	7 in. × 7 in. × 6 in.
Weight (lbs)	2	1	1	7
Power (watts)	21	14	130	100
Acoustic noise (above speech interference level)	50 dB	45 dB	68 dB	52 dB

and the vaneaxial fan both provide their high air flow rates at high pressures. Both of these fan types require over 100 W of power for their operation. The power required by a fan is approximately proportional to the product of the fan's pressure and air flow. Therefore, a fan that has one-tenth the pressure capability at a given air flow will require only one-tenth the power. For this reason the low pressure propeller and tubeaxial fans shown require much less power than the vaneaxial fan or the squirrel cage blower.

The expected life of the tubeaxial fan is 10 years when the inlet air is at room temperature. In contrast, the life of the 400 cycle vaneaxial fan is only about 5000 hours because of its high rotation speed.

The acoustic noise characteristics of the tubeaxial fan, the vaneaxial fan and the squirrel cage blower are compared in Figure 4.17. This standard method of describing the acoustic noise of fans shows the sound pressure level in decibels above 3×10^{-9} lb/in^2 as a function of frequency. The reference of 3×10^{-9} lb/in^2 is the threshold of hearing of a tone at 1000 Hz.

The "speech interference level" rating of a fan is the average noise in the three octave bands centered around 500, 1000, and 2000 Hz. The speech interference level can be determined from Figure 4.17 for the particular fans shown and has been tabulated in Table 4.1.

Low frequency acoustic noise does not sound as loud to the human ear as high frequency noise. The dotted curves of Figure 4.17 are equal loudness contours which include this effect. Note that a 37 dB pressure level at 2000 Hz "sounds" just as loud as a 40 dB level at 1000 Hz. The upper equal loudness

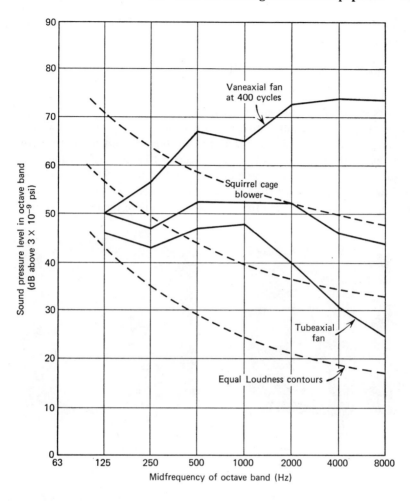

FIGURE 4.17
Comparison of the acoustic noise of tubeaxial fan, vaneaxial fan, and squirrel cage blower

contour in Figure 4.17 is the noise level in a typical business office. The lower contour is the loudness of a whisper.

The noise of the tubeaxial fan is about 10 dB below the noise level of a typical office. A special version of this fan (which can supply only 50 CFM) has its noise lowered another 20 dB to almost the level of a whisper.

The noise of the squirrel cage blower is almost as loud as the total noise of an office. The noise of the vaneaxial fan is even higher, and sounds even louder than the sound pressure level curve of Figure 4.17 suggests because the noise is concentrated at high frequencies.

Fan Accessories

Forced air cooling of electronic equipment requires the use of several accessories besides just the fan and the cooling fins.

In many designs a filter must be used to keep dust out of the electronic equipment. The filter increases the pressure that the fan must overcome to force the air through the cooling fins. Figure 4.18 shows a filter mounted on the tubeaxial fan that was shown in Figure 4.13, and also the pressure versus air flow characteristics of the fan with and without the filter. The necessity of using the filter reduces the free air flow capability of the fan from 100 CFM

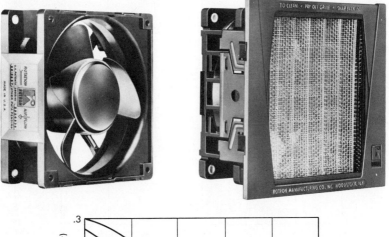

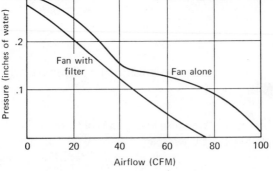

FIGURE 4.18
Vaneaxial fan with air filter and the effect of the air filter on the pressure —airflow characteristics (Photo courtesy of Rotron, Inc.)

to 75 CFM, and at 50 CFM reduces the pressure capability of the fan almost to one-half of its capability without the filter.

In some designs, the cooling fan can be mounted directly on the cooling fins. In other designs the air must be brought to the cooling fins through a supply duct. The supply duct must be chosen so that it introduces only a negligible pressure drop. This can be achieved by keeping the cross sectional area of the supply duct several times larger than the total area of the cooling ducts between the fins, and by avoiding sharp bends or changes of cross section in the supply duct.

An air flow interlock is shown in Figure 4.19. If air cooling is necessary to keep the temperature of the electronic components below their safe operating value, and if the required air flow is then not supplied because the ducting is clogged or the fan fails or is not turned on, the electronic components will be damaged. The air flow interlock, which is a pressure sensitive device actuated by the air flow itself, prevents this problem by turning off the equipment when the required flow of air is not present.

FIGURE 4.19
Airflow interlock switch (Photo courtesy of Rotron, Inc.)

4.7 EXAMPLES

The optimum design of fins for forced air cooling and the choice of the most appropriate fan can be made from the equations, figures, and tables of this chapter. The correct use of this design information is illustrated in this section by the following examples:

- The air cooled heat sink described in Section 4.1.
- A forced air cooled semiconductor heat sink.
- A forced air cooled rack of card-mounted components.
- An air cooled, high power gridded tube.

EXAMPLE 1
A Forced Air Cooled Heat Sink

The important dimensions of the forced air cooled heat sink described in Section 4.1 and shown in Figure 4.1 are as follows:

Fin spacing (w)	.090 in.
Fin height (z)	1.5 in.
Fin length (L)	1.0 in.
Fin thickness (y)	.033 in.
Fin material	Copper
Number of ducts (n)	13
Total air flow (f)	45 CFM
Inlet air temperature	20°C
Altitude	Sea level

The first step is to calculate the temperature rise of the air from Equation 4.2:

$$\frac{\Delta T \text{ (air)}}{Q} = \frac{1.73}{f} = \frac{1.73}{45 \text{ CFM}} = .04°\text{C/W}.$$

The next step is to calculate the temperature rise of the fin surface above the cooling air from Equation 4.3:

$$\frac{\Delta T \text{ (fin-air)}}{Q} = \frac{140 \, w}{n^{.2} z^{.2} f^{.8} L} = \frac{140 \times .090 \text{ in.}}{(13)^{.2} \times (1.5 \text{ in.})^{.2} \times (45 \text{ CFM})^{.8} \times 1 \text{ in.}}$$

$$= .33°\text{C/W}.$$

Equation 4.3 assumes a perfectly conducting fin. The fin temperature must be corrected to account for the finite thermal conductivity of the fin. From Equation 2.3 of Chapter 2,

$$\left(\frac{\Delta T}{Q}\right)_{\text{conduction}} = \frac{z}{kyL}$$

$$= \frac{1.5 \text{ in.}}{10 \text{ W/in. }°\text{C} \times .033 \text{ in.} \times 1.0 \text{ in.}}$$

$$= 4.5°\text{C/W}$$

where the thermal conductivity of copper from Table 2.1 of Chapter 2 is 10 W/in. °C.

$(\Delta T/Q)_{\text{transfer}}$ is .33°C/W for the heat sink of 13 fins, and is therefore 13 times greater or 4.3°C/W for a single fin.

$$\frac{\left(\dfrac{\Delta T}{Q}\right)_{\text{conduction}}}{\left(\dfrac{\Delta T}{Q}\right)_{\text{transfer}}} = \frac{4.5°\text{C/W}}{4.3°\text{C/W}} = 1.05.$$

From Figure 2.13 of Chapter 2, fin efficiency is 76%. The temperature at the base of the fin is increased by 1/.76 over the value calculated by Equation 4.3 which assumes a perfectly conducting fin.

$$\frac{\Delta T \,(\text{fin-air})}{Q} = \frac{.33°\text{C/W}}{.76} = .43°\text{C/W}.$$

The temperature difference between the tip of the fin and the air is 65% of the temperature difference between the base of the fin and the air.

These calculations can now be used to estimate the temperature at the following locations on the heat sink:

1. At the air inlet end of the heat sink at the base of the fins.
2. At the air inlet end of the heat sink at the tip of the fins.
3. At the air outlet end of the heat sink at the base of the fins.
4. At the air outlet end of the heat sink at the tip of the fins.

An air inlet temperature of 20°C and a power of 100 W will be assumed. At the inlet end of the heat sink the air temperature rise is zero because the air is just starting through the fins. The fin temperature rise is .43°C/W × 100 W = 43°C, so the fin temperature is 20°C + 43°C = 63°C at the base of the fins.

At the tip of the fins on the air inlet end, the temperature rise is 65% of the temperature rise of the base, because of the requirement to conduct heat through the fins, or 28°C. The fin temperature is therefore 48°C. At the air outlet end of the fins, the air temperature has increased by .04°C/W × 100 W = 4°C in absorbing the 100 W of power. The fin temperatures at the base and tip of the fins are correspondingly 4°C higher at the air outlet end, neglecting any axial conduction along the heat sink.

The temperature rise of the fins above the inlet air temperature at their base in the center of the heat sink from these calculations should be .43°C/W + .02°C/W = .45°C/W. Half of the total air temperature rise was taken as this would be the air temperature halfway through the fins. The measured value, as shown in Figure 4.2, is .26°C/W, and this degree of correlation is typical of the accuracy of calculations made with Equation 4.3.

TABLE 4.2
The effect of design changes on heat sink performance

Change	Units	Reference	Material	Fin Spacing	Fin Height	Fin Length	Fin Thickness
Fin spacing (w)	inches	.090	.090	.060	.090	.090	.090
Fin height (z)	inches	1.5	1.5	1.5	3.0	1.5	1.5
Fin length in direction of air flow (L)	inches	1.0	1.0	1.0	1.0	2.0	1.0
Fin thickness (y)	inches	.033	.033	.033	.033	.033	.063
Material	—	Copper	Aluminum	Copper	Copper	Copper	Copper
$\Delta T/Q$ (without fin efficiency)	°C/watt	.33	.33	.22	.29	.17	.33
Fin efficiency	—	.76	.65	.68	.60	.76	.86
$\Delta T/Q$	°C/watt	.43	.51	.32	.49	.22	.39
Pressure	inches of water	.73	.73	1.75	.18	.81	.73

Pressure drop is calculated from Equation 4.4 as follows:

$$\Delta p = \frac{\left(\frac{f^2}{n}\right)}{(wz)^2}\left[1 + .01\,\frac{L}{w}\right] \times 10^{-3}$$

$$= \frac{\left(\frac{45\ \text{CFM}}{13}\right)^2}{(.090\ \text{in.} \times 1.5\ \text{in.})^2}\left[1 + .01\,\frac{1\ \text{in.}}{.090\ \text{in.}}\right] \times 10^{-3}$$

$$= .73\ \text{in. of water.}$$

The measured value of pressure drop shown in Figure 4.3 at this flow rate of 45 CFM is 1.2 in. of water. This degree of correlation is typical of the accuracy of calculations made with Equation 4.4.

The calculated effect of varying the material of the heat sink, fin spacing, fin height, fin length, and fin thickness are shown in Table 4.2. These calculations were made in the same manner as outlined above.

The average temperature rise per watt of power and the pressure required to force air through the fins for each design change are compared on the last two lines of the table. Note that increasing fin height significantly reduces pressure requirements and that increasing fin length in the direction of air flow significantly decreases the temperature rise with only a very small increase in pressure.

EXAMPLE 2
A Forced Air Cooled Semiconductor Heat Sink

A forced air cooled heat sink with several semiconductors mounted on it is shown in Figure 4.20 along with its important dimensions. This heat sink has been carefully designed to use the tubeaxial fan that was shown in Figure 4.13, and this fan can be seen mounted on the back end of the heat sink.

The individual quadrants of the heat sink are thermally isolated, and so the calculations can be made for a single quadrant. Each quadrant uses one-fourth of the air flow supplied by the fan. The calculations will be made for an air flow of 12 CFM/quadrant, or 48 CFM total.

The first step is to calculate the temperature rise of the air above the inlet air temperature. The temperature rise of the air is calculated from Equation 4.2 as follows:

$$\frac{\Delta T\,(\text{air})}{Q} = \frac{1.73}{f} = \frac{1.73}{12\ \text{CFM}} = .14^\circ\text{C/W.}$$

Halfway through the fins, at a point directly under the component, the air temperature rise is $.07^\circ$C/W.

DIMENSIONS OF ONE QUADRANT

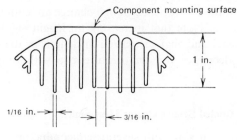

Component mounting surface

1 in.

1/16 in.   3/16 in.

Fin length = 6 in.

FIGURE 4.20
Forced air cooled semiconductor heat sink
(Photo courtesy of Wakefield Engineering,
Inc.)

The temperature rise of the fins above the cooling air is calculated from Equation 4.3. The duct dimensions are not uniform, so to use Equation 4.3 the small ducts at each end of the structure will be neglected. The number of ducts used in the calculation is therefore eight, and their average height is 1 in.

$$\frac{\Delta T \text{ (fin-air)}}{Q} = \frac{140 \ w}{n^{.2}z^{.2}f^{.8}L}$$

$$= \frac{140 \times .187 \text{ in.}}{(8)^{.2}(1 \text{ in.})^{.2}(12 \text{ CFM})^{.8} \times 6 \text{ in.}}$$

$$= .39°C/W.$$

Fin efficiency should be calculated next. The fins are made of aluminum, whose thermal conductivity from Table 2.1 of Chapter 2 is 5.5 W/in. °C. From Equation 2.3 of Chapter 2:

$$\left(\frac{\Delta T}{Q}\right)_{conduction} = \frac{z}{kyL}$$

$$= \frac{1 \text{ in.}}{5.5 \text{ W/in. °C} \times .063 \text{ in.} \times 6 \text{ in.}}$$

$$= .5°C/W.$$

$(\Delta T/Q)_{transfer}$ is .39°C/W for the heat sink of eight fins, and is therefore eight times higher for a single fin.

$$\frac{\left(\dfrac{\Delta T}{Q}\right)_{conduction}}{\left(\dfrac{\Delta T}{Q}\right)_{transfer}} = \frac{.5°C/W}{8 \times .39°C/W} = .16.$$

Fin efficiency from Figure 2.13 of Chapter 2 is .94, so the temperature rise at the base of the fins is $.39/.94 = .41°C/W$.

The total temperature rise above the inlet air temperature at the center of the heat sink is obtained from Equation 4.1 by adding the temperature rise of the air, which is .07°C/W, to the temperature rise of the fins above the air, which is .41°C/W, for a total of .48°C/W. This calculated value is in excellent agreement with the measured value of .5°C/W.

The pressure drop through the heat sink can be estimated from Equation 4.4 as follows:

$$\Delta p = \frac{\left(\dfrac{f}{n}\right)^2}{(wz)^2}\left[1 + .01\frac{L}{w}\right] \times 10^{-3}$$

$$= \frac{\left(\dfrac{12 \text{ CFM}}{8}\right)^2}{(.187 \text{ in.} \times 1 \text{ in.})^2}\left[1 + .01 \times \frac{6 \text{ in.}}{.187 \text{ in.}}\right] \times 10^{-3}$$

$$= .09 \text{ in. of water.}$$

The measured pressure drop through this forced air cooled heat sink is .1 in. of water with an air flow of 12 CFM/quadrant.

EXAMPLE 3
A Forced Air Cooled Rack of Card-Mounted Components

Figure 4.21 shows a typical rack of card-mounted components. Cooling of this rack by radiation and natural convection was considered in Example 3 of Chapter 3. In this present example, forced air cooling of the same configuration will be calculated.

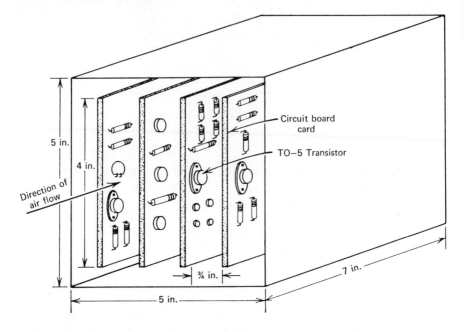

FIGURE 4.21
Forced air cooled rack of card-mounted components

The equipment shown in Figure 4.21 consists of four 4 in. × 6 in. × 1/16 in. thick circuit boards. The electronic components are mounted on one side of each of the circuit boards, and the connections between components are made with printed circuitry on the opposite side of the boards.

The four cards are mounted in a rack 5 in. wide × 7 in. long × 5 in. high. The ends of the rack are open to allow air to be blown through the cards. The spacing between the cards is approximately 3/4 in., when allowance is made for the height of the components mounted on the card surface.

The pressure versus air flow characteristics of this rack can be estimated from Curve B of Figure 4.7. At 60 CFM the pressure is about .04 in. of water. This airflow could be provided by the tubeaxial fan shown in Figure 4.18 with its filter.

An estimate of the temperature of any of the components mounted on the card can be made using Figure 4.8. A sample calculation will be made for a TO-5 transistor. The important dimensions for this calculation are:

Transistor diameter (d):   .3 in.
Cross sectional-area of duct (A):   3/4 in. × 5 in.
Airflow per duct (f):   60/6 = 10 CFM.
Height of transistor in air stream: 1/4 in.

With these dimensions

$$d\left(\frac{A}{f}\right) = \frac{.3 \text{ in.} \times 10 \text{ CFM}}{3/4 \times 5 \text{ in}^2} = .8$$

and the power transferred, as shown in Figure 4.8, is .03 W/°C for a 1 in. high component. Since the TO-5 transistor is only 1/4 in. high, the power transferred is only $1/4 \times .03$ W/°C $= .0075°$ W/°C.

If the transistor case temperature is 70°C and the air temperature is 20°C, so that the temperature difference is 50°C, a power of .4 W can be transferred by this air cooling design. This power is 6.5 times more than can be transferred with the same temperature difference by natural convection, as was shown in Example 3 of Chapter 3.

The same rack of card-mounted components will be considered again in Chapter 6, where cooling will be by liquid evaporation in a liquid filled package.

EXAMPLE 4
An Air Cooled, High Power Gridded Tube

A high power gridded tube used for industrial radio frequency heating equipment is shown in Figure 4.22 along with its important dimensions. The tube is specifically designed for forced air cooling. When the tube is operated with a plate dissipation of 3000 W, the manufacturer recommends a total air flow of 100 CFM to keep the ceramic seals below 200°C and the air cooling fins below 250°C, assuming an inlet air temperature of 50°C. At this flow rate, the measured pressure drop is .7 in. of water. The squirrel cage blower shown in Figure 4.15 is a good fan for cooling this tube.

Because the ceramic seals must be kept at a lower temperature than the cooling fins, the cooling air should be blown over the seals first and then through the fins.

The first step in calculating the forced air cooling of this tube is to calculate the temperature rise of the cooling air from Equation 4.2, as follows:

$$\frac{\Delta T \text{ (air)}}{Q} = \frac{1.9}{f} = \frac{1.9}{100 \text{ CFM}} = .019°\text{C/W}.$$

The multiplying constant of Equation 4.2 and all the other equations used in this example have been modified by the use of Figure 4.9 to account for the inlet air temperature of 50°C. When 3000 W are being absorbed, ΔT (air) $= 3000$ W $\times .019°$C/W $= 57°$C. If the inlet air temperature is 50°C, the outlet air temperature is 107°C.

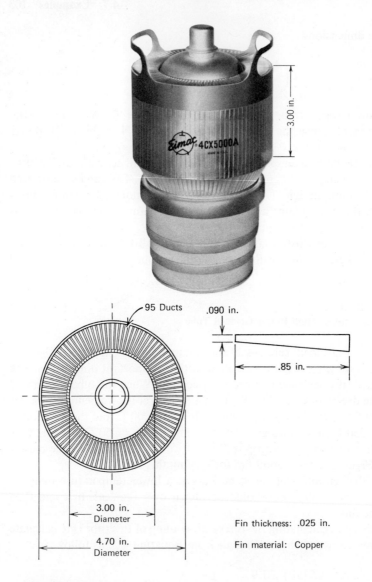

95 Ducts

.090 in.

.85 in.

3.00 in.
Diameter

4.70 in.
Diameter

3.00 in.

Fin thickness: .025 in.

Fin material: Copper

FIGURE 4.22
A forced air cooled high power gridded tube (Photo courtesy of Varian Associates)

The next step is to calculate the temperature rise of the fins above the cooling air temperature. There are 95 ducts. The shape of the ducts is not quite rectangular; the minimum dimension of .090 in. is used for the duct width. From Equation 4.3:

$$\frac{\Delta T \text{ (fin-air)}}{Q} = \frac{150 \, w}{n^{\cdot 2} z^{\cdot 2} f^{\cdot 8} L}$$

$$= \frac{150 \times .090 \text{ in.}}{(95)^{\cdot 2} \times (.85 \text{ in.})^{\cdot 2} \times (100 \text{ CFM})^{\cdot 8} \times 3 \text{ in.}}$$

$$= .047°\text{C/W.}$$

Fin efficiency should be calculated next. From Equation 2.3 of Chapter 2:

$$\left(\frac{\Delta T}{Q}\right)_{\text{conduction}} = \frac{z}{kyL}$$

$$= \frac{.85 \text{ in.}}{10 \text{ W/in. °C} \times .025 \text{ in.} \times 3 \text{ in.}}$$

$$= 1.3 °\text{C/W.}$$

$(\Delta T/Q)_{\text{transfer}}$ is .047°C/W for the total heat sink of 95 fins and is therefore 95 times greater or 4.5°C/W for a single fin.

$$\frac{\left(\dfrac{\Delta T}{Q}\right)_{\text{conduction}}}{\left(\dfrac{\Delta T}{Q}\right)_{\text{transfer}}} = \frac{1.3°\text{C/W}}{4.5°\text{C/W}} = .3$$

and from Figure 2.13 of Chapter 2, fin efficiency is 91%, and

$$\frac{\Delta T \text{ (fin-air)}}{Q} = \frac{.047°\text{C/W}}{.91} = .052°\text{C/W.}$$

When 3000 W is being transferred the temperature rise of the base of the cooling fins is 3000 W × .052°C/W = 156°C. At the inlet side of the fins the calculated fin temperature is 156°C + 50°C since the entering cooling air is at 50°C. At the outlet end of the fins the calculated fin temperature is 263°C, because the air temperature has increased 57°C in absorbing the 3000 W of power.

The pressure drop through the fins is calculated from Equation 4.5 as follows:

$$\Delta p = \frac{\left(\frac{f}{n}\right)^2}{(wz)^2} \left[1 + .01 \frac{L}{w}\right] \times .9 \times 10^{-3}$$

$$= \frac{\left(\frac{100 \text{ CFM}}{95}\right)^2}{(.090 \text{ in.} \times .85 \text{ in.})^2} \left[1 + .01 \frac{3 \text{ in.}}{.090 \text{ in.}}\right] \times .9 \times 10^{-3}$$

$$= .2 \text{ in. of water.}$$

The calculated pressure is considerably lower than the measured value of .7 in. of water, but the measured value includes the additional pressure required to force the air through the tube mounting socket.

4.8   REFERENCES

Useful references for forced air cooling of electronic equipment are as follows.

1. Krauss, A. D., *Cooling Electronic Equipment*, Prentice Hall, Inc., Englewood Cliffs, N.J., 1965, Chapter 2, pp. 30–51, Chapter 9, pp. 194–213, Chapter 12, pp. 248–269.

2. *Guide Manual of Cooling Methods for Electronic Equipment*, NAVSHIPS 900, 190, U.S. Government Printing Office, Washington, D.C., 1955, pp. 74–89.

3. *How to Select the Optimum Fan for Your Applications*, Application Bulletins 7041–7046, Pamotor, Inc., Burlingame, Calif., 1971.

4. *Fan Selection for High Altitude Applications*, Application Note B 750, Rotron, Inc., Woodstock, New York, 1969.

5. Dietz, H. G., *Forced Air Cooling Primer for the Electronics Engineer*, Henry G. Dietz Co., Long Island City, New York, 1964.

6. Sutherland, R. I., *Care and Feeding of Power Grid Tubes*, Eimac Division of Varian, San Carlos, Calif., 1967.

7. *Air Moving Devices for Commercial, Industrial and Military Applications*, Rotron, Inc., Woodstock, New York, 1970.

8. Hay, A. D., *Air Cooling Electronic Enclosures*, McLean Engineering Laboratory, Princeton Junction, N.J., 1966.

Reference 1 provides, in Chapters 2 and 12, the basic hydro-dynamic equations from which the formulas and graphs of this chapter are derived. It also provides, in Chapter 9, a detailed discussion of the theory of fan operation and the effect of ducting dimensions on air pressure drop.

Reference 2 provides the same information as Reference 1 in somewhat less detail.

Reference 3 describes the tradeoff between the factors that must be considered in fan selection, and presents a particularly good discussion of fan acoustic noise.

Reference 4 discusses the procedure for extrapolating sea level measurements to high altitude, including the choice of fan.

Reference 5 provides practical rules for designing effective forced air cooling systems.

Reference 6 provides practical rules for applying forced air cooling to high power transmitter tubes, including the scaling of sea level designs to high altitudes and high inlet air temperatures.

Reference 7 provides application information and a detailed listing of representative fans and blowers for cooling electronic equipment.

Reference 8 discusses the forced air cooling of racks and cabinets.

5

Forced Liquid Cooling

Forced liquid cooling provides more than an order of magnitude greater heat transfer per area of fin surface than forced air cooling. For very high power electronic equipment, this greater heat transfer capability is a necessity, because the cooling fin area cannot be made great enough to transfer heat by air cooling. This is particularly true when the equipment must operate at high altitudes where air density is low and forced air cooling is not very effective.

The use of liquid cooling also permits high density mounting of low power electronic components. This high density mounting may not only be desirable from a packaging standpoint; it may be a necessity to achieve required electronic performance.

The use of forced liquid cooling also eliminates the acoustic noise and vibration problems associated with forced air cooling fans. The liquid coolant pumps and heat exchangers can be removed from the electronic package, so that quiet, vibration-free operation can be maintained.

Actually, liquid cooling does not completely solve the cooling problem, it just removes it from the electronic package and puts it at another location. In the case of forced air cooling, heat is transferred to the air and then the air is exhausted to the surrounding environment and replaced with new cooling air. It is usually not practical to exhaust the cooling liquid into the surroundings, so the liquid is continuously recirculated through the electronic equipment. However, before the liquid can be recirculated, the heat that it has absorbed from the electronic equipment must be removed. This heat removal is accomplished in an air-to-liquid heat exchanger. The heat is then transferred from the fins of the heat exchanger to the surroundings by radiation and natural convection or by forced air cooling. The problems of rejecting heat to the surrounding environment has now been transferred to the air-to-liquid heat exchanger, but here it can be very effectively solved. The fins of the heat exchanger are not limited by the requirement to fit around some type of electronics package. Large noisy fans can be used because the heat exchanger is removed from the electronic components. The problem of conducting heat uniformly to the large area of the fins is solved because the liquid circulates through the fins themselves.

In designing liquid cooling for electronic equipment, the following factors must be considered:

- The design of the cooling ducts.
- The type of cooling liquid.
- The effect of coolant inlet temperature.
- The coolant pump, heat exchanger, and other auxiliary equipment.

A typical forced liquid cooling design is described in Section 5.1. The design of cooling ducts, the choice of coolant, and the effect of coolant temperature are presented in Sections 5.2 through 5.4, respectively. The necessary coolant pumps, heat exchangers, and the other auxiliary equipment required for forced liquid cooling are described in Section 5.5. Examples of forced liquid cooling designs of electronic equipment are presented in Section 5.6 to illustrate the use of the equations and figures of this chapter.

5.1   A TYPICAL FORCED LIQUID COOLING DESIGN

Figure 5.1 shows a simple liquid cooled heat sink. Four SCRs are mounted on an aluminum plate. A 3/8 in. diameter copper tubing is also mounted on the plate as shown, and cooling liquid is circulated through the tubing, which runs down one edge of the plate and back along the opposite edge.

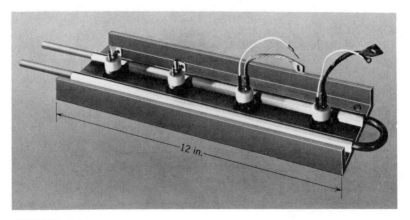

FIGURE 5.1
Forced liquid cooled heat sink (Photo courtesy of Wakefield Engineering, Inc.)

The conduction of heat from the electronic components, through the aluminum heat sink, and into the copper tubing can be calculated by the methods of Chapter 2. Care must be taken to minimize the thermal interfaces where the SCRs are mounted to the heat sink and where the tubing is mounted to the heat sink.

The temperature rise of the walls of the cooling tubing is shown as a function of coolant flow rate in Figure 5.2. This curve shows the temperature rise of the tubing walls above the inlet coolant temperature per watt of power transferred, when water is used as the coolant liquid. Flow rate is expressed in gallons per minute because liquid flow and pump characteristics are usually given in these units. The curve of Figure 5.2 has been calculated from the formulas that will be presented in Section 5.2, and the calculation is shown in Example 1 of Section 5.6.

At water flow rates of less than .2 gal/min, the flow is laminar. The liquid flows in layers along the tubing walls and does not mix between layers. Under laminar flow conditions, the temperature rise of the tubing wall required to transfer a given amount of power decreases very slowly with increasing flow rate. For example, from Figure 5.2, increasing the flow rate ten times from .02 to .2 gal/min decreases the temperature of the walls only by a factor of two times.

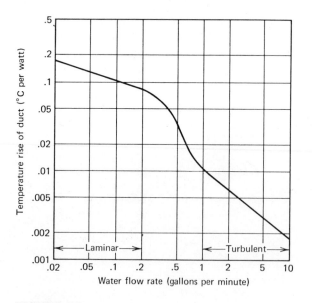

FIGURE 5.2
Temperature rise of forced liquid cooled heat sink as a function of water flow rate

As the water flow rate increases, the flow begins to be turbulent and the heat transfer capability increases significantly. Note that the temperature rise of the tubing wall decreases about eight times as the water flow rate is increased from .2 to 1 gal/min as the flow makes the transition from laminar to turbulent flow. At higher flow rates above 1 gal/min in the turbulent range the temperature rise of the duct walls is more dependent on flow rate than in the laminar flow range.

The coolant flow rate at which the flow changes from laminar to turbulent depends both on the type of coolant and on the duct geometry. Formulas for estimating this transition point are given in Section 5.2.

Air flow behaves in the same way as the liquid flow illustrated in Figure 5.2. At low air flow rates the flow is laminar, and as the air flow rate increases the flow becomes turbulent. Turbulent flow is easily achieved with forced air cooling, so only turbulent flow formulas were considered for the air cooling designs in Chapter 4. Unfortunately, with many of the high viscosity liquids commonly used to cool electronic equipment, turbulent flow cannot be achieved. Consequently, with liquid cooling, design information for both laminar and turbulent flow must be available. This information is presented in Sections 5.2 through 5.4.

At a water flow rate of 1 gal/min, the temperature rise of the cooling lines on the heat sink shown in Figure 5.1 is, as shown in Figure 5.2, .01°C/W of transferred power. If the temperature rise of the cooling duct is allowed to be 50°C above the inlet water temperature, then the heat sink is capable of transferring 5 kW of power. This is an order of magnitude more than the heat that could be transferred from the forced air cooled heat sink described in Section 4.1 of Chapter 4. The total fin area of that air cooled heat sink was 40 in^2. In contrast, the duct surface area of the water cooled heat sink is only 5/16 in. $\times \pi \times 24$ in. $= 24$ in^2, so at a flow rate of 1 gal/min the heat transfer capability per square inch of fin surface is 40 times better for the liquid cooled heat sink. Even at an extremely low water flow rate of .02 gal/min, the heat transfer capability of the liquid cooled heat sink is better than that of the air cooled heat sink.

With cooling by radiation and natural convection or by forced air, the cooling fins can operate at any temperature that will not damage the electronic component. The ultimate application of this fact was the radiation cooled high power vacuum rectifier tube described in Example 2 of Chapter 3, where the anode operated at 650°C to radiate 28 W/in^2.

In contrast, the fin temperature with forced liquid cooling is limited by the requirement that it must be lower than the boiling temperature, the decomposition temperature, or the ignition temperature of the coolant liquid. (These limitations are discussed in more detail in Section 5.4.) For example, if water is used as a coolant, the fin temperature cannot be above the boiling temperature of water (100°C), or else steam will form in the cooling lines and the required water flow cannot be maintained.

Figure 5.2 suggests that as great a coolant flow rate as possible should be used to insure that the flow is turbulent and to maximize heat transfer capability.

However, the greater the coolant flow rate, the greater the pressure required to force the coolant through the ducts. Pressure as a function of water flow rate is shown in Figure 5.3 for the particular heat sink of Figure 5.1. Pressure is expressed in pounds per square inch (psi) for liquid cooling. This choice of unit is used because pump pressure characteristics are commonly

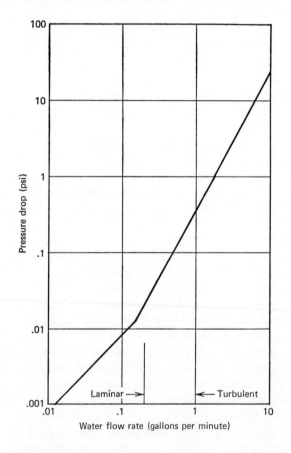

FIGURE 5.3
Pressure required to force water through the liquid cooled heat sink as a function of water flow rate

expressed in these units, and because liquid pressure gauges are normally calibrated in these units.

At water flow rates below .2 gal/min, where the flow is laminar, the pressure required to force the coolant through the duct is directly proportional to flow rate. In the turbulent flow region, pressure increases much more rapidly with increasing flow rate. Note, for example, that when the flow is increased from .2 to 10 gal/min, the pressure increases by three orders of magnitude.

Fortunately, the problem of obtaining liquid pumps with adequate pressure capability is not nearly so severe as obtaining high pressure fans for forced air cooling. Although the pressure requirements of liquid cooling designs must be carefully considered, a suitable pump can usually be found.

The effectiveness of the various methods of heat transfer—radiation and natural convection, forced air, forced liquid, and liquid evaporation—were compared in Figure 1.2 of Chapter 1.

The value for forced liquid cooling was taken from the heat sink design shown in Figure 5.1. The duct surface area is 5/16 in. $\times$ π $\times$ 24 in. = 24 in^2. At a water flow rate of 1 gal/min, the temperature rise of the duct walls is .01°C/W, as shown in Figure 5.2. Assuming an inlet water temperature of 25°C, and a fin temperature just below 100°C, so that the cooling water will not boil, the water temperature rise is 75°C. The power that can be transferred per unit area of surface is

$$\frac{75°C}{.01°C/W \times 24 \text{ in}^2} = 300 \text{ W/in}^2.$$

5.2   DESIGN EQUATIONS FOR FORCED LIQUID COOLING

The temperature of the walls of a liquid cooling duct is given by:

$$T \text{ (duct)} = \Delta T \text{ (coolant)} + \Delta T \text{ (duct)} + T \text{ (coolant inlet)} \qquad (5.1)$$

where:

T (duct)	is the temperature of the walls of the cooling duct (°C).
ΔT (coolant)	is the temperature rise of the coolant as it absorbs heat from the duct (°C).
ΔT (duct)	is the temperature of the duct walls above the coolant (°C).
T (coolant inlet)	is the inlet temperature of the coolant (°C).

The temperature rise of the duct walls above the inlet coolant temperature consists of two parts. The coolant temperature increases as the coolant flows through the ducts because the coolant is absorbing power. This is the term ΔT (coolant) of Equation 5.1.

The duct walls must be hotter than the coolant to transfer heat. This is the term ΔT (duct).

Formulas for calculating both the temperature rise of the coolant and the temperature rise of the duct walls are given in Equations 5.2 through 5.5 of Table 5.1. At first glance these equations appear rather imposing, but as the discussion of the following paragraphs will show, they are easy to apply.

Equation 5.2 gives the temperature rise of the coolant. This temperature rise depends only on the coolant flow rate and the type of coolant and is independent of duct geometry.

As illustrated in the typical liquid cooled heat sink described in Section 5.1, at low flow rates the coolant flow is laminar. The temperature rise of the duct wall above the coolant is calculated from Equation 5.4 when the flow is laminar. As shown by Equation 5.4, this temperature rise depends on flow rate, coolant thermal properties, and duct geometry. When the flow rate is increased sufficiently, the flow begins to become turbulent. The flow rate at which laminar flow ceases is given by Equation 5.3. This critical value of flow rate depends both on coolant properties and on duct dimensions and geom-

TABLE 5.1

Formulas for Forced Liquid Cooling

Temperature Rise of Coolant

$$\frac{\Delta T \text{ (coolant)}}{Q} = \frac{3.8 \times 10^{-3}}{CgF} \tag{5.2}$$

Transition Between Laminar and Turbulent Flow

$$\frac{F\phi^{.5}}{nA^{.5}} < .3\frac{\mu}{g} \text{ for laminar flow} \tag{5.3}$$

Temperature Rise of Duct Above Coolant for Laminar Flow

$$\frac{\Delta T \text{ (duct)}}{Q} = .20\frac{1}{C^{.33}K^{.67}g^{.33}}\frac{\phi^{.67}}{L^{.67}F^{.33}n^{.67}} \tag{5.4}$$

Temperature Rise of Duct Above Coolant for Turbulent Flow

$$\frac{\Delta T \text{ (duct)}}{Q} = .27\frac{\mu^{.4}}{C^{.4}K^{.6}g^{.8}}\frac{A^{.4}\phi^{.6}}{F^{.8}Ln^{.2}} \tag{5.5}$$

Pressure Drop for Laminar Flow

$$\Delta P = 6.0 \times 10^{-6}\mu\frac{FL}{nA^2\phi} \tag{5.6}$$

Pressure Drop for Turbulent Flow

$$\Delta P = 2.0 \times 10^{-5}g^{.8}\mu^{.2}\frac{F^{1.8}L}{n^{1.8}A^{2.4}\phi^{.6}} \tag{5.7}$$

TABLE 5.1 *Continued*

where:

ΔT (coolant)	= the temperature rise of the cooling liquid in absorbing heat (°C).
ΔT (duct)	= the temperature rise of the duct surface above the cooling liquid (°C).
Q	= the total power being transferred (watts).
ΔP	= the pressure drop required to force the coolant liquid through the duct (psi).
A	= cross sectional area of cooling duct (inch²).
ϕ	= dimensionless duct geometry factor equal to 4π times the cross sectional area of the duct divided by the square of its perimeter.
L	= length of duct in the direction of coolant flow (inch).
F	= total coolant flow rate $\left(\dfrac{\text{gallon}}{\text{minute}}\right)$
μ	= viscosity of coolant $\left(\dfrac{\text{pound}}{\text{hour-foot}}\right)$
K	= thermal conductivity of coolant $\left(\dfrac{\text{BTU}}{\text{hour foot}^2\text{-°F/foot}}\right)$
C	= specific heat of coolant $\left(\dfrac{\text{BTU}}{\text{pound-°F}}\right)$
g	= specific gravity of coolant relative to water.
n	= number of ducts into which coolant flow is divided.

etry. At flow rates a few times greater than this critical value, where the flow is completely turbulent, the temperature rise of the duct is calculated from Equation 5.5. This possibility of the flow being either laminar or turbulent depending on flow rate, duct size and geometry, and coolant properties is the reason that three equations instead of only one are required for the calculation of duct temperature rise.

Laminar flow can only be maintained under ideal conditions. Any obstructions, bends, or changes in cooling line cross section will cause local turbulence to occur in the flow pattern. If the flow rate is near the transition value where the flow changes from laminar to turbulent, these local perturbations will cause the entire flow to become turbulent. Equation 5.3 has been derived for the ideal case, and represents the maximum flow under which laminar flow conditions can be maintained. In many practical designs, turbulent flow will occur at lower flow rates than predicted by Equation 5.3, because of the local perturbations in the flow path.

The pressure required to force the coolant through a straight section of the cooling duct is calculated from Equation 5.6 for laminar flow or from Equation 5.7 for turbulent flow. As with the calculation for the temperature rise of the duct wall, Equation 5.3 is first used to determine the condition of the flow, and then either Equation 5.6 or 5.7 is used to calculate pressure drop.

Table 5.1 also defines all symbols that are used and gives the units in which the quantities are to be measured. Temperatures are in degrees centigrade, all dimensions are in inches, and power is in watts. Flow is in gallons per minute and pressure drop in pounds per inch^2. All these units are consistent with the measuring equipment which is normally used for electronic equipment.

The use of Equations 5.2 through 5.7 requires a knowledge of four thermal properties of the coolant: specific heat, thermal conductivity, viscosity, and specific gravity. For use in the equations, these properties should be expressed in the engineering units shown: BTUs, hours, feet, etc. This choice of units for the thermal properties has been made because the coolant supplier's data is usually presented in these units. However, should the thermal properties of a desired coolant be expressed in other units, they can be converted to the engineering units of Equations 5.2 through 5.7 by the conversion factors tabulated in Appendix B.

Equations 5.2 through 5.7 can be used for any coolant by substituting into the equations appropriate values of the thermal properties of the particular coolant. To illustrate the use of Equations 5.2 through 5.7, the thermal properties of water at an inlet temperature of 25°C are tabulated in Table 5.2. The properties of the coolant enter into Equations 5.2 through 5.7 in various combinations and to various fractional powers. The appropriate combination of these properties for each equation is also given in Table 5.2 for the case when water is used as a coolant.

The use of other liquids commonly used to cool electronic equipment is discussed in detail in Section 5.3. A similar table of the thermal properties of each of these coolants and the appropriate combination of each for direct use in Equations 5.2 through 5.7 is presented in that section. The thermal properties of coolants change with temperature, and consequently different coefficients must be used for the equations at different coolant inlet temperatures. This effect is discussed in detail in Section 5.4.

The geometry of the cooling ducts is taken into account in a way to make the equations particularly useful for electronic cooling applications. Most often the cooling duct is round in cross section, as in the liquid cooled heat sink described in Section 5.1. Equations 5.3 through 5.7 are expressed in terms of duct area and a dimensionless duct geometry factor. This geometry factor is equal to 4π times the area of the duct divided by the square of its perimeter. The usefulness of this factor comes about because it is equal to one for a round duct and, therefore, for a round duct drops out of all the equations.

TABLE 5.2

Thermal Properties of Water at 25°C

Thermal Properties

Specific heat C	1.0 BTU/lb °F
Viscosity μ	2.3 lb/hr ft
Thermal conductivity K	$.34\dfrac{\text{BTU}}{\text{hr ft}^2 \text{ °F/ft}}$
Specific gravity g	1.0

Coefficients of Equations

Equation	Symbol	Value
5.2 ΔT (coolant)	Cg	1.00
5.3 Transition: laminar to turbulent flow	$\dfrac{\mu}{g}$	2.3
5.4 ΔT (duct)/Q for laminar flow	$\dfrac{1}{C^{.33}K^{.67}g^{.33}}$	2.1
5.5 ΔT (duct)/Q for turbulent flow	$\dfrac{\mu^{.4}}{C^{.4}K^{.6}g^{.8}}$	2.6
5.6 ΔP for laminar flow	μ	2.3
5.7 ΔP for turbulent flow	$g^{.8}\mu^{.2}$	1.2

The use of Equations 5.2 through 5.7 can best be illustrated by applying them to some simple cases. This will be done in the following paragraphs.

Temperature Rise of the Coolant

The temperature rise of the coolant as it absorbs power is given by Equation 5.2. This equation, with water as a coolant, is plotted in Figure 5.4, and shows the temperature rise of the water per watt of absorbed power as a function of flow rate. Note that the temperature rise of the coolant is independent of duct size or geometry or whether the flow is laminar or turbulent. It depends only on the coolant flow rate and the specific heat and specific gravity of the coolant.

At a flow rate of 1 gal/min, the temperature rise of water is only 4×10^{-3}°C/W or 4°C/kW.

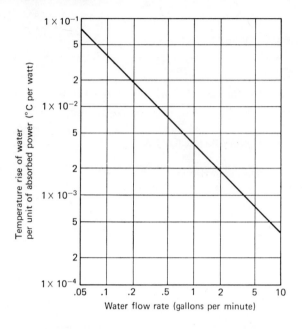

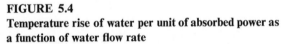

FIGURE 5.4
Temperature rise of water per unit of absorbed power as
a function of water flow rate

The Effect of Coolant Flow Rate on Duct Temperature Rise and Pressure

The effect of coolant flow rate on duct temperature rise and pressure was shown in Figures 5.2 and 5.3 for the water cooled heat sink described in Section 5.1. These figures illustrate the flow rate dependence of Equations 5.3 through 5.7. At low flow rates where the flow is laminar, the duct temperature rise varies inversely as the .33 power of the coolant flow rate, as indicated by Equation 5.4. At high flow rates where the flow is turbulent, the duct temperature rise varies inversely as the .8 power of the coolant flow rate, as shown by Equation 5.5.

This means that an eight times increase in flow rate would only reduce duct temperature to one-half when laminar flow was occurring; however, if the flow was turbulent an increase in flow rate of eight times would result in a reduction of the temperature to about one-fifth.

The pressure drop increases directly as the flow rate for laminar flow, as shown by Equation 5.6. For turbulent flow, the pressure drop increases as the 1.8 power of the flow rate, as shown by Equation 5.7.

Duct Size

The effect of duct size on the duct temperature rise and on the pressure required to force the coolant through the duct is illustrated for one particular case in Figure 5.5. These curves have been calculated from the equations of Table 5.1 for the case when the coolant is water at an inlet temperature of 25°C, the flow rate is 1/4 gal/min, and the duct is 6 in. long and has a round cross section. The temperature rise of the duct wall per unit of power transferred to the coolant is shown as a function of the diameter of the round cooling duct. Also shown is the pressure required to force the coolant through the duct (in pounds per inch²).

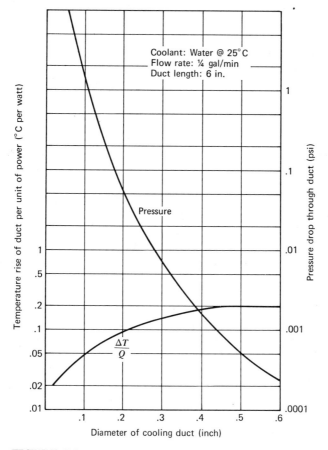

FIGURE 5.5
Temperature rise of cooling duct and pressure drop through the duct as a function of duct diameter

The temperature rise increases as the duct is made larger and larger until the duct diameter reaches about .4 in., where further increase in the duct diameter has no effect on the temperature rise. It might be expected that as the duct diameter increased, the temperature would decrease because the area for heat transfer has increased; however, as the duct diameter increases the flow characteristics change, and this compensates for the increase in heat transfer area. The point where the temperature rise no longer changes with duct diameter represents the transition from turbulent flow with small ducts to laminar flow with large ducts. The fact that this transition occurs at a duct diameter of .4 in. is not universal, but depends on the particular type of coolant and flow rate chosen. As Equation 5.5 shows, the temperature rise depends on the .4 power of the duct area for turbulent flow, and as Equation 5.4 shows, it is independent of the duct area for laminar flow. These facts are illustrated in Figure 5.5. The curve of Figure 5.5 suggests that as the diameter of the cooling duct is made smaller and smaller, the temperature rise can be reduced to arbitrarily small values. Although this is true, it is important to consider the pressure required to force the flow through the duct, and this is also shown in Figure 5.5. As the diameter is changed from .025 in. to .600 in., the temperature rise of the duct changes only by one order of magnitude. In contrast, the pressure changes about six orders of magnitude. As Equation 5.7 shows, the pressure drop for turbulent flow depends on the duct area inversely as the 2.4 power of the duct area, and as Equation 5.6 shows, the pressure drop has about the same dependence for laminar flow, varying inversely as the 2.0 power of the area. Consequently, as Figure 5.5 illustrates, there is an extremely rapid variation in pressure with duct diameter.

Duct Geometry

The discussion of the previous paragraph and the example of the liquid cooled heat sink described in Section 5.1 considered only ducts with a round cross section. Equations 5.3 through 5.7 include the shape of the duct by means of a dimensionless "duct geometry factor." This factor is equal to 4π times the area of the duct divided by the square of its perimeter. The factor is chosen in this form so that it is equal to one for a duct with a round cross section and so drops out of the equations.

The difference in cooling effectiveness between a round and a rectangular duct is compared in Figure 5.6. This figure is calculated using Equations 5.3 through 5.7 and shows the temperature rise of the duct wall per unit of power and the pressure drop through the duct as functions of coolant flow rate.

The coolant is water at an inlet temperature of 25°C and the duct length is 3 in. Both the round and the rectangular duct have the same cross-sectional

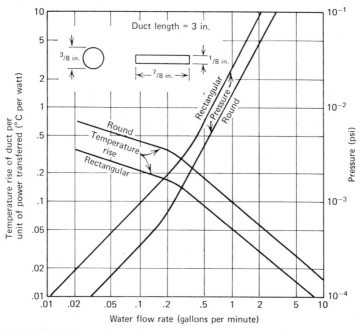

FIGURE 5.6
Effect of the shape of the duct cross section on duct temperature rise and pressure

area. The inner diameter of the round duct is 3/8 in. and the rectangular duct is 7/8 in. × 1/8 in. The dimensionless duct geometry factor is one for the round duct but is only .35 for the rectangular duct. The effect of the rectangular geometry is to reduce the temperature rise of the duct walls by about a factor of two and to increase the pressure drop by about the same factor.

The temperature rise of the round duct could of course be decreased by the same factor of two simply by reducing its diameter, but the pressure drop would be increased much more than two times. From Figure 5.5., if the diameter of the round duct were reduced to .080 in., the temperature rise would be the same as for the 7/8 in. × 1/8 in. rectangular duct, but the pressure drop would not be increased just two times, but by a factor of about 20 times.

Duct Length

It would be expected that if the duct length were doubled and the same total power dissipated, the duct wall temperature rise would be reduced by one-half. This is true when the flow is turbulent, as is shown by Equation 5.5.

However, in the laminar flow case, the temperature rise of the duct wall varies inversely as the duct length to the .67 power, as shown by Equation 5.4, because of the nature of the buildup of laminar layers along the duct wall. This effect is illustrated in Figure 5.7, which shows the duct temperature rise per unit of power as a function of flow rate for two different duct lengths, 3 in. and 12 in. The coolant is water at an inlet temperature of 25°C. The curve for the 3 in. long duct is the same as that shown in Figure 5.6. Note that the improvement in heat transfer with the longer duct is less when the flow is laminar than when it is turbulent.

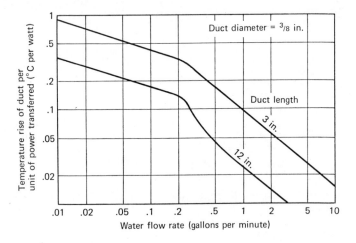

FIGURE 5.7
Effect of duct length on duct temperature rise and pressure

Multiple Ducts

The heat transfer capability with a given flow of coolant can sometimes be enhanced by dividing the flow up into a number of parallel ducts.

For laminar flow, as shown by Equation 5.4, the temperature rise of the duct walls per watt of power varies inversely as $n^{.67}$, where n is the number of ducts into which the total coolant flow F is divided. For example, if the flow is divided into ten ducts, the temperature rise of the duct walls is reduced by a factor of five.

For turbulent flow, as shown by Equation 5.5, the temperature of the duct walls per unit of power varies inversely as $n^{.2}$. For example, if the flow is divided into ten ducts, the temperature will be reduced only by a factor of 1.6.

Dividing the flow into multiple ducts can significantly reduce the pressure required to force the total amount of coolant through the equipment.

The pressure drop depends inversely on n for laminar flow and inversely on $n^{1.8}$ for turbulent flow, as shown by Equations 5.6 and 5.7 respectively.

With turbulent flow, the cooling effectiveness and the pressure both depend on duct diameter. Therefore, when the coolant is divided into n ducts, the duct diameter can be reduced to improve cooling without increasing the pressure over the single duct case. If Equations 5.5 and 5.7 are combined to eliminate the duct area at constant pressure, the temperature rise of the duct walls per total power transferred can be shown to vary inversely as $n^{.5}$ for turbulent flow.

The Effect of Bends and Fittings on Pressure Drop

The formulas for calculating pressure drop given in Table 5.1 and illustrated in the figures of this section were for uniform straight lengths of coolant duct. Any bends, changes in duct dimensions, or fittings add greatly to the total pressure drop through the lines. In most liquid cooled electronic equipment, the major source of pressure drop is the bends and fittings in the cooling line, and not the drop through the straight lengths of duct. (This fact has already been mentioned in connection with Equation 4.4 of Chapter 4 which describes the pressure drop through forced air cooling fins. In that case the pressure drop through the fins themselves was comparable to the pressure drop of the entrance and exit transitions only when the fin length was greater than 100 times the fin spacing.)

For this reason, an exact calculation of pressure drop through the system is extremely difficult. An approximate result may be obtained by assigning an effective length to each transition such that the pressure drop through a straight duct with this length is the same as the pressure drop of the transition. Figure 5.8 shows effective lengths of the most common bends and transitions. The use of Figure 5.8 is best illustrated by a simple example of a 24 in. long cooling line with a 90% bend in the center. The cooling line has a round cross section with an inner diameter of 5/16 in. From Figure 5.8, the ratio of the effective length of the 90° bend to the duct diameter is 30, so the effective length of the bend is 30 × 5/16 in. = 10 in. or 40% of the total duct length. The pressure drop through the coolant tubing is therefore 1.40 times higher than would be calculated from the Equations 5.6 or 5.7 which are for straight lengths of duct. If the coolant lines contain several bends or transitions, the pressure drop can be several times greater than would be calculated for a straight duct.

The effective length factors of Figure 5.8 apply strictly to turbulent flow cases, but they can be used for a rough estimate in laminar flow cases. In either case, a cooling design obtained by the use of the Equations of Table 5.1 and Figure 5.8 must be built and experimentally evaluated. Suitable

EFFECTIVE LENGTH OF TRANSITIONS FOR PRESSURE DROP CALCULATIONS

Type of transition	Sketch	Effective length*
90° Square elbow	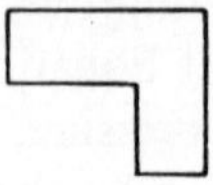	60
90° Round elbow	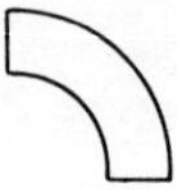	30
180° Round bend	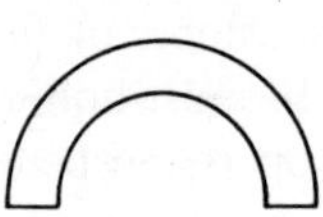	75
Contraction or enlargement ratio = ¾	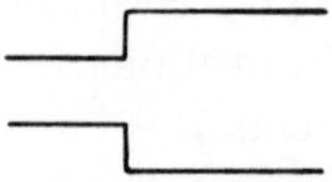	10
Contraction or enlargement ratio = ½	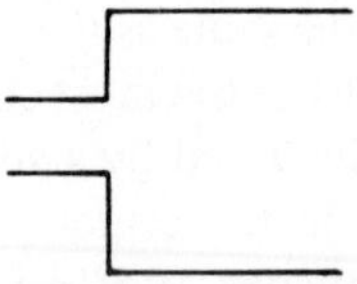	25

*Effective length is expressed as number of duct diameters

FIGURE 5.8
"Effective length" of transitions for pressure drop calculations

measurement techniques for forced liquid cooling designs are described in detail in Chapter 10.

A reasonable value of total pressure drop through liquid cooled electronic equipment is 25 psi. If the water flow is 1 gal/min, and the coolant duct through the electronic equipment has a round cross section with an inner diameter of 5/16 in., the use of Equation 5.7 shows that the pressure drop is

.012 psi/in. of effective duct length. Therefore, the effective duct length, which takes into account all bends and fittings, can be 2000 in. or approximately 160 ft long for a pressure drop of 25 psi. As shown in Section 5.3, the pressure drop with high viscosity hydraulic oils may be an order of magnitude greater than with water. In either case, the problem of designing the coolant ducts to minimize pressure drop is much less severe with liquid cooling than with forced air cooling.

5.3 COOLANT LIQUIDS

The choice of coolant liquid is determined by the following factors:

1. Heat transfer capability.
2. Low temperature limit of operation.
3. High temperature limit of operation.
4. Insulating properties.
5. Ability to perform hydraulic functions as well as cooling.
6. Inertness, compatability with seals and tubing.
7. Flammability, toxicity.

Often the coolant liquid with the best heat transfer properties cannot be used, because of the need to satisfy the other of the above requirements. For example, water has excellent heat transfer capabilities, but if the electronic equipment must be operated below the freezing temperature of water, water cannot be used. Antifreeze can be added to the water to lower its freezing temperature, but the heat transfer capability of the water-antifreeze mixture is poorer than that of pure water.

If the coolant liquid must be circulated through several electronic components which are at different voltages, then the insulating properties of the coolant must be considered. In this case, neither water nor the water-anti-freeze mixture would be satisfactory and an oil would have to be used. The best oil from a heat transfer standpoint would be a fluorochemical. However, if the coolant liquid has to serve the dual function of both a coolant and a hydraulic fluid, another type of oil would have to be used.

Once the coolant liquid has been chosen to meet all the equipment requirements listed above, the cooling ducts must be designed to effectively use the coolant to cool the equipment. Equations 5.2 through 5.7 can be used for this design. These equations can be used for any coolant by substituting appropriate values of the thermal properties of the particular coolant.

Electronic equipment is commonly cooled with water, water and antifreeze, fluorochemical or silicone dielectric oils, or hydraulic fluids. To provide a convenient reference, and to illustrate the effect of coolant liquid selection

TABLE 5.3
Thermal Properties of Coolant Fluids

Coolant	Temperature of coolant	Thermal Property				Coefficient of Equation					
		Specific heat, C	Viscosity μ	Thermal Conductivity, K	Specific gravity, g	5.2 ΔT(Coolant) $1/Cg$	5.3 Transition, μ/g	5.4 ΔT(Duct)/Q Laminar $\dfrac{1}{C^{.33}K^{.67}g^{.33}}$	5.5 ΔT(Duct)/Q Turbulent $\dfrac{\mu^{.4}}{C^{.4}K^{.6}g^{.8}}$	5.6 ΔP Laminar μ	5.7 ΔP Turbulent $g^{.8}\mu^{.2}$
Water	25°C	1.00	2.3	.34	1.0	1.00	2.3	2.1	2.6	2.3	1.2
	100°C	1.00	0.6	.39	1.0	1.00	0.6	1.9	1.4	0.6	0.9
Antifreeze-Water*	25°C	.74	10.0	.22	1.1	1.25	9.3	3.0	8.4	10.0	1.7
	100°C	.83	2.0	.20	1.0	1.18	2.0	3.0	4.2	2.0	1.2
FC-75	25°C	.25	3.5	.08	1.8	2.28	1.9	7.1	8.1	3.5	2.1
	100°C	.28	1.3	.07	1.5	2.38	0.9	7.9	6.7	1.3	2.3
Coolanol 45	−25°C	.40	420	.085	.93	2.70	450	7.3	76	420	3.2
	25°C	.45	41	.082	.89	2.50	46	7.3	30	41	1.9
	100°C	.52	7.8	.075	.86	2.86	9.1	7.5	16	7.8	1.3
Air at Sea Level	20°C	.25	.044	.015	.0012	3300	36.7	240	1300	.044	.0026

* 62% Ethylene glycol—32% water

on heat transfer capability, the thermal properties of a typical coolant of each type are tabulated in Table 5.3.

The following liquids are compared in this table:

1. *Water:* Water is the best liquid from a cooling standpoint. However, it suffers from the limitations of a low boiling temperature, freezing at temperatures below 0°C, and poor insulating capabilities.

2. *A Mixture of 62% Ethylene Glycol and 38% Water:* This is the commonly used antifreeze mixture and will not freeze until −65°C is reached, which is the lower limit of military requirements.

3. *FC-75:* This is a fluorochemical coolant which has excellent dielectric properties. It has the best coolant properties of any available dielectric fluid. It is usuable over the temperature range from −65°C to 100°C.

4. *Coolanol 45:* This is a commonly used hydraulic fluid. It makes possible the use of a single fluid to provide all fluid functions in an electronic system, since it can serve the function of coolant fluid, hydraulic fluid, and dielectric fluid. Because it serves all these functions, it serves none of them as well as special purpose fluids and, consequently, it is not as good a coolant as water, water-ethylene glycol or FC-75. It can be used over the temperature range from −65°C to 175°C, although its cooling properties vary widely over this range.

5. *Air:* The thermal properties of air at sea level are included for comparison. The forced air cooling equations of Chapter 4 were derived using the air thermal properties listed in Table 5.3 and Equations 5.2 and 5.5 and 5.7 for turbulent flow.

The properties of each coolant enter into Equations 5.2 through 5.7 in various combinations and to various fractional powers. The appropriate combination of these properties for each equation is also given in Table 5.3.

The thermal properties of these commonly used liquid coolants change with coolant temperature. Therefore, the thermal properties and coefficients of the equations are tabulated for each coolant at several temperatures. The effect of this temperature sensitivity is discussed in detail in Section 5.4.

The cooling effectiveness of the various liquids can best be compared by some simple examples. These comparisons are discussed in the following paragraphs.

Temperature Rise of the Coolant

The temperature rise of each of the four typical coolants—water, water and ethylene glycol, FC-75, and Coolanol 45—are compared in Figure 5.9. The figure shows the temperature rise per watt of absorbed power as a function of coolant flow rate.

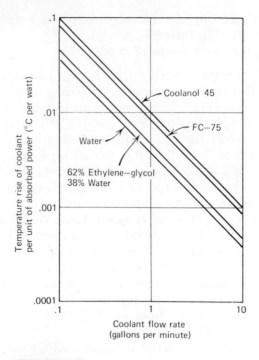

FIGURE 5.9
Temperature rise of the coolant as a function of flow rate for four different coolants

A particularly interesting point to compare the curves is at a flow rate of 1 gal/min, which is typical of the flow available, in many electronic systems. If water is used, the temperature rise is only 4×10^{-3}°C/W or 4°C/kW. If antifreeze is added to the water, the temperature rise is about 5°C/kW. If FC-75 or Coolanol 45 is used, the temperature rise is 10°C/kW. Consequently, as far as the rise in temperature of the coolant is concerned, even the worst fluid considered is only one-third as poor a coolant as water. Quite a different situation occurs when the temperature rise of the duct above the coolant is considered, as will be shown in the next paragraph.

The Effect of Coolant on Duct Temperature Rise and Pressure

The effects of coolant selection on the temperature rise of the duct wall and on the pressure drop are shown in Figures 5.10 and 5.11, respectively. These curves are calculated from Equations 5.3 through 5.7 using the thermal properties of each coolant at 25°C from Table 5.3. For all calculations a straight duct 3 in. long with a round cross section 3/8 in. in diameter is assumed.

Figure 5.10 shows the temperature rise of the duct wall per unit of power transferred as a function of coolant flow rate. As can be seen, water is the best coolant, with ethylene glycol and water, FC-75, and Coolanol 45 following in that order. When all the fluids are flowing under turbulent flow conditions, for example, at about a flow of 10 gal/min, the ethylene glycol-water mixture or FC-75 gives about three times the temperature rise that would be obtained with water. The Coolanol 45 gives about ten times the temperature rise. Consequently, the difference between water and the poorest coolant, as far as the temperature rise of the duct is concerned, is a factor of ten, whereas when considering the temperature rise of the coolant, as discussed in the previous paragraph, the difference between water and the worst fluid is only a factor of three.

The flow at which the transition between laminar and turbulent flow occurs depends primarily on the viscosity of the coolant fluid. As shown in Table 5.3, the viscosity varies from 2.3 for water to 41 for the Coolanol 45. Therefore, water enters the turbulent flow region at a much lower flow rate than does Coolanol 45. This fact is illustrated in Figure 5.10.

In the laminar flow region, the difference in the coolant effectiveness of the various fluids is much less than it is in the turbulent flow region. For

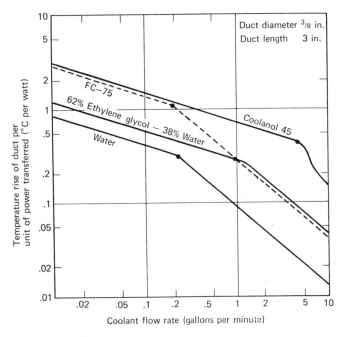

FIGURE 5.10
Temperature rise of the duct as a function of flow rate for four different coolants

example, at a flow rate of about .02 gal/min, the worst coolant, Coolanol 45, gives a duct temperature rise only three times that of water, in contrast to the case of turbulent flow where the ratio was ten times.

FC-75 is a particularly interesting fluid. It has a specific heat and a thermal conductivity comparable to that of the hydraulic oils, but a viscosity comparable to that of water. Consequently, under laminar flow FC-75 behaves exactly like the hydraulic oils, whereas under turbulent flow it behaves exactly like the water-ethylene glycol mixture.

The pressure drop of the various coolant fluids as a function of flow rate is shown in Figure 5.11. When all the fluids are in turbulent flow, as is achieved at a flow of 10 gal/min, there is very little difference in the pressure; as a matter of fact, the difference is less than two to one.

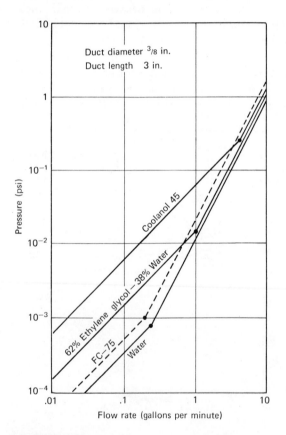

FIGURE 5.11
Pressure required to force the coolant through the duct as a function of flow rate for four different coolants

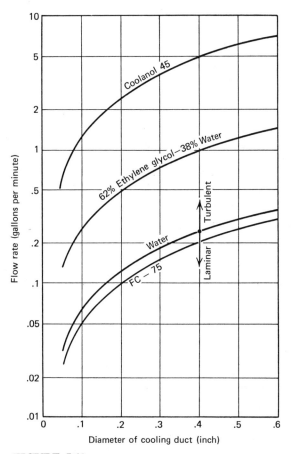

Diameter of cooling duct (inch)

FIGURE 5.12
Flow rate at which the flow changes from laminar to turbulent as a function of cooling duct diameter for four different coolants

In contrast, for laminar flow, such as occurs for all the fluids at about .05 gal/min, there is an order of magnitude difference in pressure between water and Coolanol 45.

The major difference between the coolant liquids in regards to their thermal properties is their viscosities, and this causes the major difference between their performance as coolants. The pressure drop depends only slightly on viscosity for turbulent flow; consequently, the pressure is approximately independent of the choice of liquid. In contrast, the temperature rise of the duct wall depends strongly on viscosity and therefore on the choice of coolant for turbulent flow. Just the opposite is true for laminar flow. The temperature rise of the duct for laminar flow does not depend on viscosity, so there is not so much difference between coolants, but the pressure drop for laminar flow does depend on the viscosity, and, hence, there is a great difference in the pressure drop between coolants.

The flow rate and cooling duct diameter at which the transition between laminar and turbulent flow occurs is shown in Figure 5.12 for the four fluids being compared. For any combination of flow rate and duct diameter lying under the curve of a particular fluid, the flow is laminar. For any combination of duct diameter and flow rate lying above the curve for the fluid, the flow is turbulent. As stated previously, turbulent flow occurs at large flow rates and small duct diameters, whereas laminar flow occurs at larger duct diameters and lower flow rates. The particular value of diameter and flow rate at which the transition occurs for a given fluid depends primarily on the viscosity of the fluid, and for the fluids considered there is a wide variation of viscosity. At a flow rate of 1 gal/min and duct diameters larger than about .100 in., the hydraulic oils will be in laminar flow and water, ethylene glycol-water, and the FC-75 will always be in turbulent flow. Water, ethylene glycol-water, and FC-75 will only be in laminar flow if the flow rate is very small, for example, less than about 1/10 gal/min.

5.4   COOLANT TEMPERATURE

The problems of coolant temperature that must be considered when designing forced liquid cooling for electronic equipment are:

- Minimum operating temperature.
- Maximum coolant temperature.
- The effect of coolant temperature on heat transfer and pressure drop.

These problems are discussed in this section.

Minimum Operating Temperature

The major problem of low coolant temperature is getting the coolant liquid to flow. Water freezes and therefore will not flow at temperatures below 0°C, unless antifreeze has been added. The fluorochemical and hydraulic oils do not freeze, but their viscosity increases as the temperature is lowered so that they do not flow. The critical low temperature for these dielectric fluids is their "pour" point, which is the lowest temperature at which the liquid will flow. Obviously, the liquid cannot be used below this temperature, but even at temperatures above this point, the viscosity of the oil may be so high that the pump does not have sufficient pressure capability to pump the required volume of liquid through the cooling system. The viscosities of the

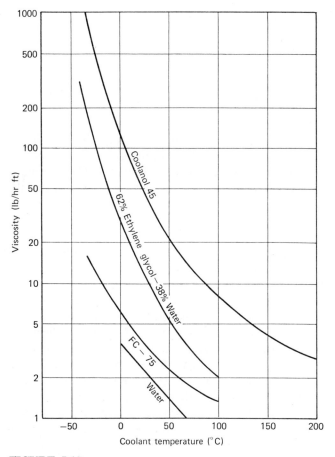

Coolant temperature (°C)

FIGURE 5.13
Viscosity of four different coolants as a function of coolant temperature

four coolants discussed in the previous section—water, FC-75, a mixture of
62% ethylene glycol with water, and Coolanol 45—are shown as a function
of temperature in Figure 5.13.

The pressure drop when the coolant is in laminar flow was given by
Equation 5.6 of Table 5.1, and is directly proportional to viscosity. As Figure
5.13 shows, over the temperature range from 25°C to -25°C, the viscosity
of Coolanol 45 or the ethylene glycol-water mixture changes by approximately an order of magnitude. Consequently, the pressure required to force
a given flow rate through the cooling system changes by an order of magnitude
over this temperature range.

Maximum Coolant Temperature

The problems associated with high coolant temperatures are:

- Boiling of the coolant.
- Decomposition.
- Ignition.

The walls of the coolant duct must be hotter than the cooling fluid in order to transfer heat to the coolant. Formulas for the temperature rise of the duct above the coolant were given in Table 5.1. These formulas give the coolant duct temperature relative to the average temperature of the coolant. Actually a temperature gradient exists through the fluid. Near the duct wall, the coolant is at approximately the same temperature as the duct wall. Therefore, the duct wall temperature must be kept below the safe operating temperature of the coolant, because the coolant near the duct wall reaches approximately the duct wall temperature.

The temperature of the coolant near the duct wall must not be allowed to exceed the boiling temperature of the coolant. As will be discussed in Chapter 6, a high rate of heat transfer can be achieved by allowing the coolant to boil at the surface of an electronic component. However, to effectively use this technique, adequate means must be provided to allow the vapor formed during the boiling process to escape. If the wall of the duct used for forced liquid cooling exceeds the boiling temperature of the liquid, the liquid will boil and the pressure inside the duct will increase so that the required liquid flow rate cannot be maintained. Consequently, boiling must be prevented in forced liquid cooling designs, and this must be achieved by limiting the duct wall temperature to below the boiling temperature of the liquid.

Another factor that limits the maximum allowable temperature of the fluid is decomposition. Hydraulic oils will decompose at high temperature before they boil, and they will clog the coolant ducts with carbon deposits so that the required amount of coolant can no longer be circulated through the ducts.

The third problem that must be considered with regard to the maximum temperature of the coolant is ignition. When coolants are raised above their "flash" point, vapor will be generated which will ignite when the liquid is brought in contact with a spark or other ignition means. Consequently, the coolant temperature at the duct wall must be maintained below the flash point of the liquid.

The boiling temperature, decomposition temperature, and flash point of the four coolants discussed in the previous section are compared in Table 5.4.

TABLE 5.4

Maximum Coolant Temperatures

Coolant	Boiling Temperature	Decomposition Temperature	"Flash" Point
Water	100°C	—	—
62% Ethylene glycol-38% water	100°C	—	—
FC-75	100°C	450°C	—
Coolanol 45	—	180°C	375°C

The Effect of Coolant Temperature on Heat Transfer and Pressure Drop

The maximum and minimum temperatures at which the various coolants can be used have been discussed in the previous paragraphs. Within the temperature range set by these limits, the thermal properties of the coolant change considerably and consequently their cooling effectiveness and the pressure required to force them through the cooling duct changes. Values of the thermal properties of the various coolants were shown at several temperatures in Table 5.3. As this table shows, the thermal property that is most sensitive to coolant temperature is viscosity and the wide variation of viscosity with temperature was shown graphically in Figure 5.13.

The effect of coolant inlet temperature on the temperature rise of the coolant duct is shown in Figure 5.14 for the case of Coolanol 45 flowing through a 3 in. long round duct with a 3/8 in. inner diameter. Curves are shown for three different values of average coolant temperature: −25°C, 25°C, and 100°C. The curve for a coolant temperature of 25°C is identical to that shown in Figure 5.10, where the heat transfer characteristics of various coolant liquids were compared at 25°C. At this temperature, the coolant flow is laminar until a flow rate of approximately 5 gal/min is reached. If the coolant temperature is lowered to −25°C, because of the higher viscosity of the fluid at this low temperature, the flow remains laminar until a flow rate of 40 gal/min is reached. In most liquid cooled electronic systems, this high flow rate would not be used, so that over the entire range of practical coolant flow rates, the flow will always be laminar when the average temperature of the Coolanol 45 is −25°C.

In contrast, when the temperature of the Coolanol 45 is raised to 100°C, the transition from laminar to turbulent flow occurs at a much lower flow rate of 1 gal/min.

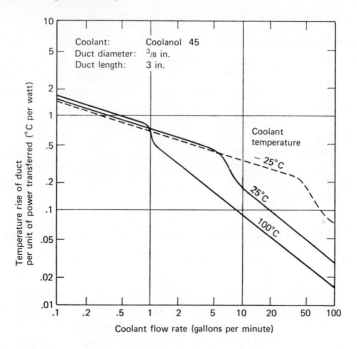

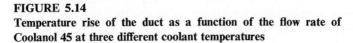

FIGURE 5.14
Temperature rise of the duct as a function of the flow rate of Coolanol 45 at three different coolant temperatures

In a typical electronic equipment cooled with Coolanol 45, the flow rate might be 1 gal/min. Over the full range of temperature from −25°C to 100°C the flow would be laminar, and as shown by Figure 5.14, there would be a negligible difference of duct temperature rise with coolant inlet temperature.

However, over this same inlet temperature range the pressure required to force the coolant through the duct varies greatly. This variation of pressure with coolant inlet temperature is illustrated in Figure 5.15, where the pressure required to force the Coolanol 45 through the same 3 in. long 3/8 in. diameter duct is shown as a function of flow rate for the three coolant inlet temperatures: −25°C, 25°C, and 100°C. At a flow rate of 1 gal/min, in the laminar flow region, the pressure varies almost two orders of magnitude over the coolant inlet temperature range. At −25°C, the pressure drop is almost 1 psi, and this decreases to .01 psi as the coolant inlet temperature is raised to 100°C.

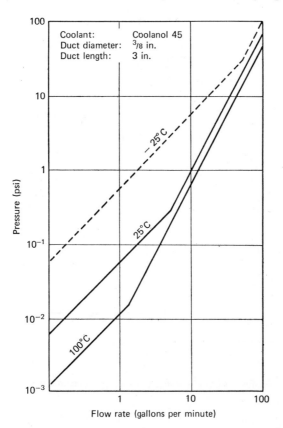

FIGURE 5.15
Pressure required to force Coolanol 45 through a
duct as a function of flow rate at three different
coolant temperatures

5.5 HEAT EXCHANGERS, COOLANT PUMPS, AND AUXILIARY EQUIPMENT

A block diagram of the equipment required for forced liquid cooling is shown in Figure 5.16. This equipment includes:

- Coolant pump.
- Air to liquid heat exchanger.
- Coolant reservoir.
- Coolant lines.
- Pressure interlock.
- Flow interlock.

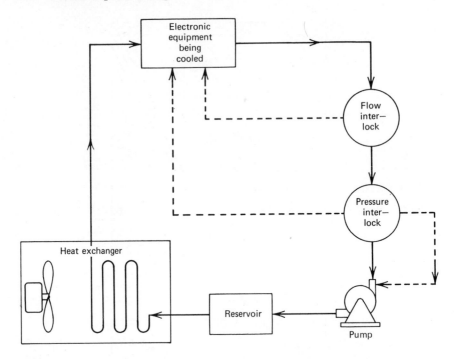

FIGURE 5.16
Block diagram of a forced liquid cooling system for electronic equipment

The block diagram clearly indicates that forced liquid cooling is much more complicated than forced air cooling which requires only a fan.

A complete packaged forced liquid cooling unit is shown in Figure 5.17. This packaged unit contains all the elements of the block diagram of Figure 5.16. The coolant pump, reservoir, and air to liquid heat exchanger with its cooling fan can be seen in the photograph of the cooling unit with the cover removed. The particular unit shown is designed to circulate 1.5 gal/min of water at pressures up to 50 psi. With this water flow rate, the unit can transfer 10 kW of power from the water to the surroundings. The unit requires 11 A at 110 V of 60 cycle AC power or 1.2 kW of power for its operation. It weighs 75 lb and fits into a standard 19 in. relay rack. The dimensions of the front panel of the unit are 17 in. × 14 in.

One major advantage of forced liquid cooling is that the complete cooling unit, including the heat exchanger, pump, reservoir, and all auxiliary equipment can be removed from the electronic package. In this way, noise and vibration can be eliminated from the electronic equipment. The air-to-liquid heat exchanger and the coolant pump are not limited in design by the require-

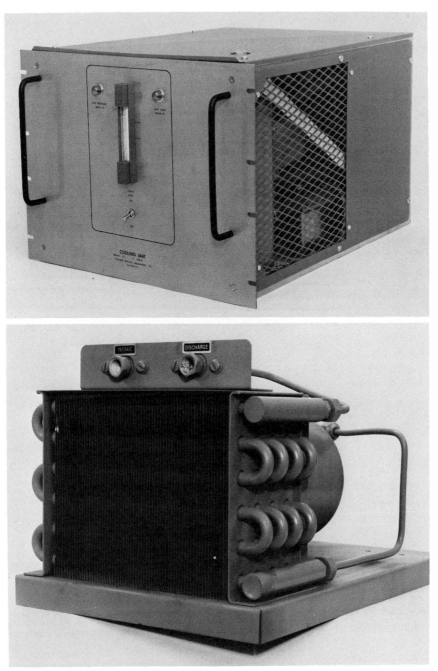

FIGURE 5.17
Typical forced liquid cooling system for use with electronic equipment (Photo courtesy of Electro Impulse Laboratory, Inc.)

ment of fitting around the electronic equipment. Because the pumps and heat exchangers are not limited by any electronic requirements, these items can best be designed by heat transfer specialists. The electronic engineer can concentrate his cooling design on the coolant ducts in the electronic equipment, where a tradeoff between electrical and cooling performance must be achieved, and then purchase the necessary auxiliary equipment to provide the coolant liquid as a complete cooling unit such as shown in Figure 5.17.

The auxiliary equipment required for a forced liquid cooling system includes a pressure interlock switch and a flow interlock switch. The purpose of the pressure interlock switch is primarily to protect the pump and heat exchanger from damage caused by excessive pressure. This condition might occur if the connecting hoses became pinched or if any blockage occurred in the coolant system. When the pressure in the coolant lines rises above a preset level, the pressure interlock is activated and turns off both the coolant pump (which prevents damage to the coolant circulation system) and the electronic component (so that it is not damaged by overheating when the coolant stops circulating).

The flow interlock is used primarily to prevent damage to the electronic component if the coolant flow rate drops below the level required to adequately cool the equipment. If this occurs, the flow interlock is actuated and the power to the electronic equipment is removed.

5.6 EXAMPLES

Forced liquid cooling of electronic equipment can be designed from the equations, figures, and tables of this chapter. The correct use of this design information is illustrated in this section by the following examples:

- The water cooled SCR heat sink described in Section 5.1.
- A water cooled high power microwave transmitter tube.
- Water cooling of the microwave cavities of the high power transmitter tube.
- A high power microwave transmitter tube cooled with hydraulic oil.

EXAMPLE 1
A Water Cooled SCR Heat Sink

As shown by Equation 1.1 of Chapter 1, the temperature rise of an electronic component above the surroundings consists of two parts:

1. The temperature rise required to *conduct* heat from the component to the cooling lines.
2. The temperature rise required to *transfer* heat from the cooling lines to the surroundings.

The water cooled heat sink described in Section 5.1 and shown in Figure 5.1 illustrates both of these parts.

Figure 5.18 shows the important dimensions of this heat sink, and also shows the heat flow paths. Calculations will be made for a water flow rate of 1 gal/min. Each SCR will be assumed to dissipate 150 W, so the total power is 600 W.

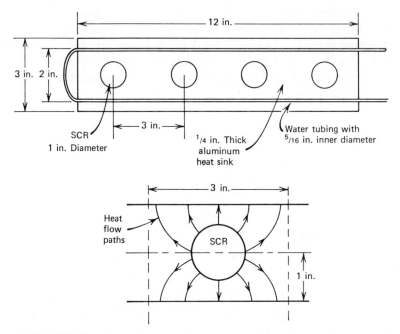

FIGURE 5.18
Dimensions of liquid cooled heat sink and heat flow paths

The first step is to calculate the temperature rise of the water. This temperature rise can be determined directly from Figure 5.4, or can be calculated from Equation 5.2.

From Figure 5.4, which shows the temperature rise of water per watt of absorbed power as a function of water flow rate:

$$\frac{\Delta T \text{ (coolant)}}{Q} = 3.8 \times 10^{-3}\,^\circ\text{C/W}$$

at a flow rate of 1 gal/min. Alternately, using Equation 5.2 with the appropriate thermal properties of water from Table 5.2:

$$\frac{\Delta T \text{ (coolant)}}{Q} = \frac{3.8 \times 10^{-3}}{CgF} = \frac{3.8 \times 10^{-3}}{1 \times 1 \text{ gal/min}} = 3.8 \times 10^{-3}\,^\circ\text{C/W}.$$

The dissipated power is 600 W and so the temperature rise of the coolant is $\times 3.8\,10^{-3}\,°\text{C/W} \times 600\,\text{W} = 2\,°\text{C}$.

The next step is to calculate the temperature rise of the duct walls above the coolant. For these calculations the equations of Table 5.1 are used, with the thermal properties of water taken from Table 5.2. First, a decision must be made as to whether the flow is laminar or turbulent, using Equation 5.3. This can be determined immediately from Figure 5.12 where Equation 5.2 is plotted to show the transition flow rate for various liquids as a function of cooling duct diameter. The point for 1 gal/min flow rate and .312 in. duct diameter lies above the transition curve for water, so the flow is turbulent.

The temperature rise of the duct can next be calculated, using Equation 5.5 for turbulent flow:

$$\frac{\Delta T \,(\text{duct})}{Q} = .27\,\frac{\mu^{.4}}{C^{.4}K^{.6}g^{.8}}\,\frac{A^{.4}\phi^{.6}}{F^{.8}Ln^{.2}}$$

$$= \frac{.27 \times 2.6 \times (.076\,\text{in}^2)^{.4} \times (1)^{.6}}{(1\,\text{gal/min})^{.8} \times 24\,\text{in.} \times (1)^{.2}}$$

$$= .01\,°\text{C/W}$$

where:

$(\mu^{.4}/C^{.4}K^{.6}g^{.8})$ for water is 2.6 from Table 5.2.

A (the area of the duct) is $(\pi/4)(5/16)^2 = .076\,\text{in}^2$.

ϕ (the duct geometry factor) is 1.

F (the flow rate) is 1 gal/min.

L (the duct length) is 24 in.

n (the number of ducts) is 1.

The calculated temperature rise of the duct walls per watt of power transferred was shown in Figure 5.2 for this particular heat sink. The calculations were made as shown above.

The total power dissipated in the heat sink is 600 W, so the temperature rise of the duct walls is

$$.01\,°\text{C/W} \times 600\,\text{W} = 6\,°\text{C}.$$

The next step is to calculate the temperature rise required to conduct heat from the base of each SCR to the cooling lines. Assuming the heat flow geometry shown in Figure 5.18, Equation 2.3 can be used

$$\Delta T = \frac{Q\lambda}{k\alpha}$$

$$= \frac{75\,\text{W} \times 1\,\text{in.}}{5.5\,\text{W/in.}\,°\text{C} \times 3\,\text{in.} \times 1/4\,\text{in.}}$$

$$= 18\,°\text{C}$$

where:

Q is the power conducted toward one side of the heat sink (half the power dissipated by each SCR) and is 75 W.

λ is the effective length through which heat is conducted and is 1 in.

α is the effective cross section through which heat is conducted and is $3 \times 1/4$ in².

k is the thermal conductivity of aluminum and is 5.5 W/in. °C.

The next step is to determine the temperature difference required to conduct heat across the interface between each SCR and the heat sink. A typical value can be obtained from Figure 2.11 of Chapter 2. Assuming heat sink compound is used and the clamping torque is 300 lb in., the thermal resistance is .10°C/W. The total power dissipated by each SCR is 150 W, so the temperature difference is 15°C.

The total temperature difference between the SCR and the cooling liquid can now be calculated.

	Temperature Difference
Interface between SCR and heat sink	15°C
Conduction through heat sink	18°C
Temperature rise of duct above coolant	6°C
Temperature rise of coolant	2°C
Total Temperature Difference	41°C

If the inlet water is at 35°C, the SCR temperature is 76°C. Note that most of the temperature difference is required to conduct the heat from the SCR to the cooling line, rather than to transfer it from the cooling line to the water.

The final step is to calculate the pressure drop required to force 1 gal/min of water through the cooling line. From Equation 5.7 for turbulent flow:

$$\Delta P = 2.0 \times 10^{-5} (g^{.8} \mu^{.2}) \frac{F^{1.8} L}{n^{1.8} A^{2.4} \phi^{.6}}$$

$$= \frac{2.0 \times 10^{-5} \times 1.2 \times (1 \text{ gal/min})^{1.8} \times 48 \text{ in.}}{(1)^{1.8} \times (.076 \text{ in}^2)^{2.4} \times (1)^{.6}}$$

$$= .6 \text{ psi.}$$

All symbols and values are as defined previously, except for the length of coolant lines, which must now include the effect of the 180° bend. From

Figure 5.8, the ratio of the effective length of the 180° bend to the duct diameter is 75, so the effective length of the bend is 75 × 5/16 in. = 24 in. or as much as the physical length of the entire duct. The total effective length is therefore 24 in. + 24 in. = 48 in.

EXAMPLE 2
A Water Cooled, High Power Microwave Transmitter Tube

Figure 5.19 shows a water cooled high power klystron, which generates 2 kW of microwave power. Two parts of this klystron tube require liquid cooling: the collector, where 5 kW of power are dissipated, and the cavities and output window where 500 W are dissipated. A calculation of the cooling of the collector with water is made in this example. Cooling of the cavities and output window with water is calculated in Example 3. Cooling of the collector with hydraulic oil at several inlet oil temperatures is discussed in Example 4.

FIGURE 5.19
Water cooled high power klystron (Photo courtesy of Varian Associates)

A cutaway photograph of the liquid cooled collector of this klystron is shown in Figure 5.20. A sketch of the coolant flow path is also shown. The coolant enters through an inlet fitting at the top end of the collector and divides to flow through the 24 ducts in the inner body of the collector. Each duct is .062 in. × .062 in. × 3 in. long. The fins are .062 in. thick. The entire collector is made of copper. At the bottom end of the collector the coolant leaves the ducts, turns around, and flows back through the outer annular ring. The main function of this annular region is to provide a return path for the water. Some cooling is provided by the water flow in the annular region, but this will be neglected in this calculation.

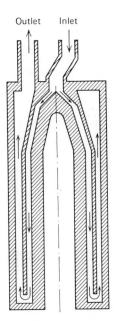

Outlet   Inlet

FIGURE 5.20
Cutaway photograph of the liquid cooled collector of the high power klystron and a schematic drawing showing liquid flow (Photo courtesy of Varian Associates)

A water flow rate of 1.5 gal/min is required to cool this collector. A suitable cooling system to supply this flow would be the unit shown in Figure 5.17, which includes a pump, reservoir, and air-to-liquid heat exchanger. With a flow rate of 1.5 gal/min, this cooling unit can maintain the water temperature at the collector inlet at 40°C, if the surroundings are at 20°C. This temperature difference of 20°C is required to transfer the absorbed 5 kW of power to the cooling air in the air-to-liquid heat exchanger.

The first step in designing the water cooled collector is to calculate the temperature rise of the water in absorbing the 5 kW of power. This temperature rise can be determined directly from Figure 5.4 (which shows the temperature rise of the water per watt of absorbed power as a function of water flow rate) to be 2.5×10^{-3}°C/W $\times$ 5 kW = 13°C. The water enters the klystron collector at 40°C and rises 13°C in absorbing the 5 kW of power. It therefore leaves the collector at 53°C and enters the heat exchanger at this same temperature of 53°C.

The 5 kW of power is removed from the water in the air-to-liquid heat exchanger, and the water temperature is lowered back to 40°C, and the water is ready for recirculation through the collector.

The next step is to calculate the heat rise of the collector cooling ducts above the cooling water. For these calculations the equations of Table 5.1 are used, with the thermal properties of water taken from Table 5.2.

A decision must first be made as to whether the flow is laminar or turbulent, using Equation 5.3.

$$\frac{F\phi^{.5}}{nA^{.5}} \text{ must be less than } .3\frac{\mu}{g} \quad \text{ for laminar flow}$$

$$\frac{F\phi^{.5}}{nA^{.5}} = \frac{1.5 \text{ gal/min } (.78)^{.5}}{24(.0040 \text{ in}^2)^{.5}} = .88$$

and

$$.3\frac{\mu}{g} = .3 \times 2.3 = .7$$

where:

F (the total water flow rate) is 1.5 gal/min.

n (the number of ducts) is 24.

A (the cross-sectional area of each duct) is .063 in. × .063 in. = .0040 in^2.

ϕ (the duct geometry factor) = $(4\pi \times .0040)/(4 \times .063)^2 = .78$.

μ/g (the coefficient of Equation 5.3 for water taken from Table 5.2) is 2.3.

The flow rate is above the limit for laminar flow, so the turbulent flow formula of Equation 5.5 will be used. Actually, the flow is in the transition region between laminar and turbulent, so the results obtained using the turbulent flow formulas may give somewhat optimistic results.

From Equation 5.5:

$$\frac{\Delta T \text{ (duct)}}{Q} = .27 \frac{\mu^{.4}}{C^{.4}K^{.6}g^{.8}} \frac{A^{.4}\phi^{.6}}{F^{.8}Ln^{.2}}$$

$$= \frac{.27 \times 2.6 \times (.004 \text{ in}^2)^{.4} \times (.78)^{.6}}{(1.5 \text{ gal/min})^{.8} \times 3 \text{ in.} \times (24)^{.2}}$$

$$= .008°C/W$$

where:

$\mu^{.4}/C^{.4}K^{.6}g^{.8}$ (the coefficient of Equation 5.5 for water taken from Table 5.2) is 2.6.

L (the length of the cooling ducts) is 3 in.

F, A, n, ϕ are as defined previously.

Equation 5.5 assumes a perfectly conducting fin. The fin temperature must be corrected to account for the finite thermal conductivity of the fin. From Equation 2.3 of Chapter 2,

$$\left(\frac{\Delta T}{Q}\right)_{conduction} = \frac{\lambda}{k\alpha}$$

$$= \frac{.062 \text{ in.}}{10 \text{ W/in. }°C \times .062 \text{ in.} \times 3 \text{ in.}}$$

$$= .033°C/W$$

where:

λ	(the height of the fins) is .062 in.
α	(the cross-sectional area of the fins) is .062 in. × 3 in.
k	(the thermal conductivity of copper) is 10 W/in. °C.

$(\Delta T/Q)_{transfer}$ is .008°C/W for the total collector of 24 fins, and is therefore 24 times greater or .19°C/W for a single fin.

$$\frac{\left(\dfrac{\Delta T}{Q}\right)_{conduction}}{\left(\dfrac{\Delta T}{Q}\right)_{transfer}} = \frac{.033°C/W}{.19°C/W} = .17$$

and from Figure 2.13 of Chapter 2, fin efficiency is 95%.

$$\frac{\Delta T \text{ (duct)}}{Q} \text{ is therefore } \frac{.008°C/W}{.95} = .009°C/W.$$

The temperature rise of the collector ducts above the cooling water is therefore. 009°C/W × 5000 W = 45°C. The collector temperature can now be calculated from Equation 5.1.

Coolant inlet temperature	40°C
Temperature rise of coolant	13°C
Temperature rise of duct above coolant	45°C
Collector Temperature	98°C

In this design, the collector surface remains, as required, below the boiling temperature of the water.

The pressure drop through the collector ducts may be estimated from Equation 5.7 for turbulent flow.

$$\Delta P = 2.0 \times 10^{-5}(g^{.8}\mu^{.2}) \frac{F^{1.8}L}{n^{1.8}A^{2.4}\phi^{.6}}$$

$$= \frac{2.0 \times 10^{-5} \times 1.18 \times (1.5 \text{ gal/min})^{1.8}[3 \text{ in.} + (120 \times .063 \text{ in.})]}{(24)^{1.8}(.004 \text{ in}^2)^{2.4} \times (.78)^{.6}}$$

$$= 1.1 \text{ psi}$$

where:

F, A, ϕ are defined previously.

$g^{.8}\mu^{.2}$ (the coefficient of Equation 5.7 for water) is 1.18.

L is the effective length including two right angle bends of equiva-
lent length $60 \times .063$ in. where the flow turns around and
enters the annular region. The effective length of each right
angle bend is taken from Figure 5.8.

This pressure drop calculation neglects the pressure drop through the
annular region and the effect of the inlet and outlet fittings and the transitions
between the fittings and the collector ducts. For an accurate determination
of the pressure drop, measurements must be made using the techniques
described in Chapter 10.

EXAMPLE 3
**Water Cooling of the Cavity and Output Window of the High Power
Microwave Transmitter Tube**

The high power klystron shown in Figure 5.19 requires liquid cooling of both
its collector (where most of the power is dissipated) and also of its cavity and
output window. The design of water cooling for the collector was discussed
in the previous example. This example will discuss the cooling of the cavity
and output window, where 500 W must be dissipated. The inlet fitting for the
liquid cooling of these parts of the klystron can be seen in Figure 5.19.
Figure 5.21 shows a sketch of the liquid cooling duct running around the
cavity and around the output window. The duct is round in cross section
with a 3/16 in. inner diameter. The total length of the cooling duct is 12 in.
The water flow rate is 1/4 gal/min.

The temperature rise of the water and the temperature rise of the cooling
duct above the water temperature can be calculated using the formulas of
Table 5.1 and the thermal properties of water given in Table 5.2, in the same
manner as the calculations of the previous two examples were made. However, for this particular case, Figures 5.4 and 5.5 can provide the answers
directly. (Figure 5.5 has been calculated for the same flow rate as used in this
example.)

The first step is to determine the temperature rise of the cooling water.
This can be determined directly from Figure 5.4. At a flow rate of 1/4 gal/min,

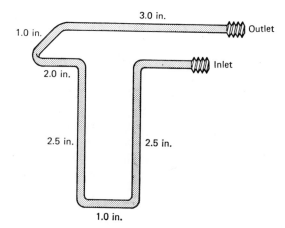

3.0 in.

1.0 in.

Outlet

2.0 in.

Inlet

2.5 in.

2.5 in.

1.0 in.

FIGURE 5.21
Schematic drawing of the liquid cooling duct around
the cavity and output window of the high power klystron

the temperature rise of the water is $1.5 \times 10^{-2} °C/W$. Since 500 W must be absorbed, the total temperature rise of the water is $1.5 \times 10^{-2} °C/W \times 500$ W $= 7.5°C$.

The next step is to determine the temperature rise of the duct walls, which can be obtained directly from Figure 5.5 which shows the temperature rise of the duct walls per unit of power as a function of the diameter of the cooling duct when the water flow is 1/4 gal/min. For a 3/16 in. diameter duct, Figure 5.5 gives the temperature rise of the duct walls of .09°C/W for a 6 in. long duct. With this flow of water and this diameter, Figure 5.5 also shows that the flow is turbulent. In this example, the duct length is 12 in. and since the temperature rise is inversely proportional to the duct length for turbulent flow, the temperature rise of the walls is .045°C/W $\times$ 500 W $= 22.5°C$ Therefore, the total temperature rise of the duct walls above the inlet water temperature, including both the temperature rise of the water and the temperature rise of the duct walls, is $7.5°C + 22.5°C = 30°C$.

In the previous example it was suggested that the cooling water for the collector of this water cooled klystron could be supplied by a cooling unit such as was shown in Figure 5.17. This unit could also supply the extra 1/4 gal/min of water required for the cavity and window cooling of the tube. In this case, the coolant would flow in parallel through the tube collector and through the cavity cooling duct. A flow regulator would have to be put in the collector cooling line to adjust the pressure so that the total flow would divide correctly. When the cooling unit shown in Figure 5.17 is used, the inlet water temperature is 40°C and so the maximum temperature of the cavities and window would be 40°C + 30°C = 70°C.

The pressure drop through the cavity and window cooling duct can be estimated using Figures 5.5 and 5.8. From Figure 5.5 with a duct diameter of 3/16 in., the pressure is approximately .1 psi for a 6 in. length of duct. The actual duct length through the klystron body is 12 in. However, as shown in the sketch of Figure 5.21, it consists of six right angle bends. Each of these bends adds greatly to the effective length that must be used in estimating pressure drop. From Figure 5.8 each of these bends adds a length equal to 60 times the diameter of the tubing, which is 60 times 3/16 in. = 11 in. Therefore, the six right angle bends add effective length of 66 in. so that the total effective length for calculating pressure drop is 78 in. Therefore, the pressure drop is approximately $78/6 \times 0.1 = 1.3$ psi.

EXAMPLE 4
A High Power Microwave Transmitter Tube Cooled with Hydraulic Oil

Water is not a suitable coolant for many airborne electronic equipments. These equipments must operate over wide extremes of temperature, where water would freeze at the low temperature extreme and boil at the high temperature extreme. A more suitable coolant in these airborne equipments is hydraulic oil, such as Coolanol 45. The oil serves both as the coolant and the hydraulic fluid for the system.

The cooling of the collector of a high power microwave tube with Coolanol 45 will be discussed in this example. Calculations will be made at three inlet oil temperatures: −25°C, 25°C, and 100°C, as these are typical of the range of coolant temperatures over which the transmitter tube must operate in an airborne radar.

For comparison purposes, the collector design will be the same as on the water cooled klystron discussed in Example 2, which was shown in Figure 5.20. Because Coolanol 45 is a poorer coolant than water, this collector design can only dissipate 1 kW when hydraulic oil is used as the coolant. The coolant flow rate will be 1.5 gal/min, the same as the water flow rate in Example 2.

The calculations are made in the same way as in Example 2, using the equations from Table 5.1 and the properties of Coolanol 45 at three temperatures from Table 5.3.

The results of the calculations are summarized in the Table 5.5. The appropriate formula used for each calculation is shown in the second column.

The first step, as shown on line one of the table, is to calculate the temperature rise of the coolant.

The next step is to determine whether the flow is laminar or turbulent, and the results of the calculations that determine this are shown on lines three, four, and five. At all coolant inlet temperatures the flow is laminar.

TABLE 5.5
Calculated Performance of Klystron Collector Cooled with Coolanol 45

Power: 1000 W
Coolant: Coolanol 45
Flow Rate: 1.5 gal/min.

	Equation	Inlet Oil Temperature		
		$-25°C$	$25°C$	$100°C$
1. Temperature rise of coolant	5.2	7°C	6°C	6°C
2. Laminar flow limit	Right side of 5.3	135	13.8	2.7
3. $\frac{F}{n}\left(\frac{A}{\phi}\right)^{.5}$ in one of the 24 ducts	Left side of 5.3	.88	.88	.88
4. Type of flow	—	Laminar	Laminar	Laminar
5. $\frac{\Delta T}{Q}$ for 24 ducts	5.4	.061°C/W	.061°C/W	.062°C/W
6. Temperature rise of ducts	—	61°C	61°C	62°C
7. Temperature of collector	5.1	43°C	92°C	168°C
8. Pressure drop thru ducts and turnaround	5.6	133 psi	13 psi	2.5 psi

The third step is to calculate the temperature rise of the cooling duct walls above the coolant and the results are shown on lines five and six of the table. This temperature rise is practically independent of coolant inlet temperature. The total temperature of the collector, which depends on the coolant inlet temperature, coolant temperature rise, and duct temperature rise, is calculated next, as shown in line seven. At the maximum inlet temperature of 100°C, the collector temperature remains below the decomposition temperature of the oil, as given in Table 5.4, of 180°C.

The final step is to calculate the pressure drop through the 24 cooling ducts and the turnaround at the end of the ducts. The results of this calculation are shown on line eight. The pressure required to force the Coolanol 45 through the ducts is 55 times higher at $-25°C$ than at $100°C$. The pressure drops shown on line eight do not include the effects of the annular return duct, the inlet and outlet fittings, or the transitions between the fittings and the ducts. Because these parts of the collector have no effect on cooling, they can usually be designed for minimum pressure drop.

The major problem of the operation of the collector with Coolanol 45 over a wide range of inlet temperatures is the pressure drop. The cooling is adequate at a flow rate of 1.5 gal/min at all inlet temperatures, but the pump may not be able to supply this required flow at the low inlet oil temperature because of the high pressure required.

5.7   REFERENCES

Useful references for forced liquid cooling of electronic equipment are as follows.

1. Krauss, A. D., *Cooling Electronic Equipment*, Prentice Hall, Inc., Englewood Cliffs, N.J., 1965, Chapter 2, pp. 30–43, Chapter 7, pp. 138–172, Chapter 13, pp. 270–285.

2. McAdams, W. H., *Heat Transmission*, McGraw Hill, New York, 1954, Chapter 9, pp. 202–251.

3. Sigel, L. A., "Coolant Selection for Electronic Systems," *Electronic Packaging and Production Magazine*, Jan., 1965, pp. 75–77 and Feb., 1965, pp. 105–107.

4. *Introduction to Cooling Units*, Technical Bulletin 696, Electro Impulse, Inc., Red Bank, N.J., 1969.

Reference 1 provides the basic hydrodynamic equations in Chapter 2 from which the formulas and graphs of this chapter are derived. Chapter 7 describes air-to-liquid heat exchangers and Chapter 13 describes liquid cooling of electronic components.

Reference 2 is another source for the basic hydrodynamic equations used to derive the formulas of this chapter.

Reference 3 classifies the various coolant liquids that can be considered for electronic equipment and compares their cooling effectiveness.

Reference 4 describes the design and application of typical cooling units to be used to supply the coolant for forced liquid cooled electronic equipment.

6

Cooling by
Liquid Evaporation

Cooling by liquid evaporation offers many advantages over other heat transfer methods:

1. Liquid evaporation provides the greatest rate of heat transfer of any of the heat transfer means, up to approximately 800 W/in^2.
2. All of the components in an electronic equipment can be simply and effectively cooled by immersing the entire equipment in a package filled with dielectric oil, and using cooling by evaporation to transfer the heat from the components to the walls of the equipment package.
3. Evaporation cooling can be used as a means of temperature stabilization of electronic components, because liquid boiling occurs at a constant temperature.
4. Very simple expendable cooling systems are made possible by evaporation cooling, by simply evaporating the cooling liquid and exhausting the resulting vapor to the surroundings.

Four typical designs of evaporation cooling of electronic equipment, which individually illustrate the above advantages, are described in Section 6.1.

Evaporation cooling is accomplished by immersing the electronic component or components to be cooled in a liquid bath. Heat is transferred from the component by boiling the liquid at the component surface.

Figure 6.1 shows a high power transmitter tube being cooled by evaporation. The liquid is evaporated at the surface of the tube anode, and the vapor bubbles rise through the liquid bath. The vapor generated by the boiling process can be exhausted from the system, but it is usually condensed back to the liquid phase in a condenser and returned to the cooling bath. The heat

153

FIGURE 6.1
An evaporation cooled transmitter tube showing the boiling heat transfer
process (Photo courtesy of ITT)

transfer area of the condenser can be made large enough, since it is not
limited by electronic design requirements, that heat can be transferred from
it to the surroundings by the simplest heat transfer means—such as forced
air cooling or even radiation and natural convection.

Factors that must be considered in designing evaporation cooling of
electronic equipment include:

1. Choice of the cooling fluid.
2. Design of the component surface where evaporation occurs.
3. Design of the condenser.
4. Pressure and pressure equalization.
5. Equipment orientation.

These design considerations are discussed in detail in Section 6.2. Special
cooling system considerations and auxiliary equipment for the evaporation

cooling of high power electronic components are described in Section 6.3. The design of liquid filled electronic packages is discussed in Section 6.4. The design of expendable cooling systems is considered in Section 6.5.

Examples of evaporation cooling of electronic equipment are presented in Section 6.6 to illustrate the use of the design information of this chapter.

6.1   TYPICAL EVAPORATION COOLING DESIGNS

Evaporation cooling can be effectively used in the following different ways in electronic equipment:

1. Cooling of high power components at high power densities.
2. Cooling of all components in an electronic equipment by immersing the entire assembly in a package filled with dielectric oil.
3. Maintaining a constant temperature bath for electronic components.
4. Simple expendable cooling systems.

Typical cooling designs which illustrate these uses of evaporation cooling are described in this section.

Evaporation Cooling of High Power Electronic Components

Figure 6.2 shows a schematic diagram of a typical evaporation cooling design for a high power transmitter tube. The anode of the tube is immersed in a bath of cooling water in the boiler section of the system. Heat is transferred from the anode by boiling the water, as was shown in Figure 6.1. The vapor (steam) resulting from the boiling process rises through the bath, into the space above the bath, and then into the condenser. The vapor condenses back to liquid in the condenser and returns to the boiler to repeat the process. The heat transferred to the condenser as the steam condenses back to water must be continuously removed by radiation and natural convection, forced air cooling, or forced liquid cooling.

The use of evaporation cooling does not completely solve the cooling problem—it just moves it from the electronic component to another location where it can be more effectively solved. The condenser size is not limited by electrical design considerations, so it can be made large enough that heat can be transferred to the surroundings by the simplest heat transfer means.

The amount of power that can be transferred from the electronic component (in watts per inch^2 of component surface) is shown in Figure 6.3 as a function of component's surface temperature, when the coolant is water at atmosphere pressure. At low surface temperatures, when the component temperature is below the boiling temperature of the water, heat is transferred by natural convection. The liquid near the component surface is heated and

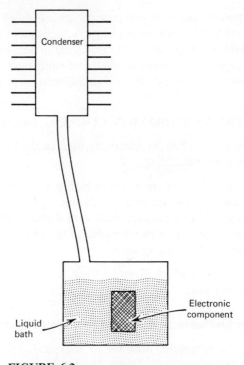

FIGURE 6.2
An evaporation cooling system for electronic equipment

rises and is replaced by cooler liquid. The convective liquid current provides the heat transfer. This natural convection process is the same as occurs with natural convection air cooling which was discussed in detail in Chapter 3.

As the temperature of the component surface increases, the amount of heat that is transferred by natural convection continues to increase until the temperature of component surface reaches a few degrees above the boiling point of the water. At this temperature of 108°C, the convective heat transfer rate is 30 W/in².

When the surface temperature of the component reaches 108°C, evaporation cooling begins. In the range from 108°C to 125°C, nucleate boiling occurs. Individual bubbles of vapor are formed at the hot component surface as the water is boiled. These bubbles of vapor break away from the surface and travel upward through the water bath. A photograph of this nucleate boiling process was shown in Figure 6.1. As the temperature of the component surface rises above 108°C, the heat transfer rate rapidly increases, until it reaches 870 W/in² at 125°C, as shown in Figure 6.3.

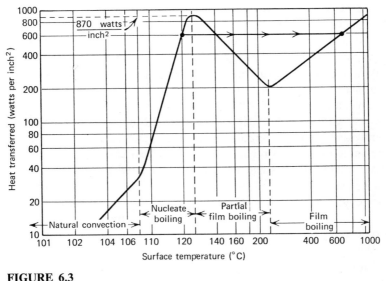

FIGURE 6.3
Heat transferred per unit area of component surface as a function of surface temperature for evaporation cooling with water

If the temperature of the component is increased above 125°C, as would occur if the power density were increased to above the maximum limit of 870 W/in^2, the heat transfer suddenly begins to decrease. This decrease in heat transfer occurs because the surface of the component becomes partially insulated by the vapor and the liquid cannot get to the surface to evaporate. This effect becomes worse and worse as the temperature of the surface increases further, and the heat transfer capability continues to decrease until a temperature of 225°C is reached, where the component surface is completely insulated by vapor. At this temperature the heat transfer has dropped to 125 W/in^2 which is one-seventh of the maximum heat transfer rate that was obtained before the vapor film began to insulate the surface.

Above 225°C, the component surface is completely insulated by a film, and as the component temperature continues to increase, the heat transfer capability again begins to increase. This range of temperature operation is called the "film vaporization" region. Note that the heat transfer capability does not again reach 870 W/in^2 until the surface reaches a temperature of approximately 1000°C.

The heat transfer versus component surface temperature curve shown in Figure 6.3 is irreversible. Assume, for example, that the surface of the component is operating at a power density of 600 W/in^2. As shown in Figure 6.3, the component temperature will be 120°C. If the power density is momentarily increased to be greater than the maximum of 870 W/in^2, the temperature of the component surface will rapidly rise above the range where nucleate boiling can occur, and the anode surface will become completely insulated by a vapor film. If the power density is then lowered back to the normal 600 W/in^2 operating range, the vapor film will remain over the surface, and the surface temperature will remain at 600°C, as shown by the arrows of Figure 6.3, and the power will be transferred by the film vaporation process. Of course within a short time the electronic component would be permanently damaged by operation at this high temperature.

The value of the maximum power density that can be transferred by the nucleate boiling process and the temperature at which this maximum heat transfer is achieved depend on the type of cooling liquid, the geometry and surface conditions of the component surface, and the pressure at which the system operates. The effects of all these factors are discussed in detail in Section 6.2.

In this typical design, evaporation cooling serves the same function as forced liquid cooling in transferring the heat from the electronic component to another heat exchanger where the heat can be more effectively transferred to the surroundings. However, in contrast to the forced liquid cooling case, no pumps are required to circulate the liquid, and much greater power densities can be transferred from the surface of the electronic component.

Although evaporation cooling offers the advantages of simplicity and high heat transfer capability, it has disadvantages also. Electronic equipment that is cooled by evaporation can be operated in one orientation only, with the heat generating elements at the lowest point in the system. Also, immediate destruction of the electronic component occurs if the maximum heat transfer rate is ever exceeded.

Liquid Filled Electronic Packages

Figure 6.4 shows another typical evaporation cooling design for electronic equipment. In this case a complete electronic assembly, consisting of vacuum tubes, solid state devices, transformers, capacitors, and resistors is completely immersed in a dielectric liquid. Dielectric liquid is used instead of water because of its insulating properties, so that the various electronic components can all operate at different voltages. Heat is transferred from all the low power components in the equipment by natural convection. These components operate at a temperature less than the boiling temperature of the dielectric liquid. Heat is transferred from the high power components by evaporation cooling. The surfaces of the high power components are a few degrees hotter than the boiling temperature of the dielectric liquid, and the liquid evaporates at the surface of these components.

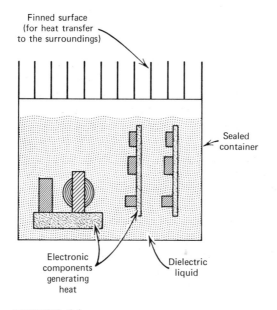

FIGURE 6.4
A liquid filled electronics package

The power density at the surface of the high power components must be kept below the critical value so that the vapor film cannot form to insulate the surface. This maximum power density for typical dielectric fluids is much lower than for water, and is approximately 100 to 250 W/in². The vapor from the boiling process rises through the liquid bath and condenses on the cooler surface of the package. The heat must then be removed from the package surface by radiation and natural convection, forced air cooling or forced liquid cooling.

The advantages of the liquid filled electronics package are:

1. All components can be effectively and simply cooled.
2. The components can operate at different voltages, without the danger of high voltage breakdown.
3. High rates of heat transfer per unit area of component surface can be obtained.

The choice of coolant fluid, the design of the component surface, the design of the condenser, and the effect of internal pressure on cooling performance are discussed in Section 6.2. Special cooling system considerations for liquid filled electronics packages, including designs to accommodate the thermal expansion of the cooling fluid when the equipment is operated over wide temperature ranges, are discussed in Section 6.4.

Constant Temperature Baths

A third electronic equipment design using evaporation cooling is shown in Figure 6.5. In this case evaporation cooling is used to provide a constant temperature environment for a klystron oscillator. The oscillation frequency of the klystron is determined by the dimensions of its tuning cavity. These dimensions can be maintained constant, independent of the temperature of the surroundings, by using evaporation cooling. As shown in the schematic drawing of Figure 6.5, the klystron is mounted in a bath of water. The klystron has been designed to transfer all of the heat through its resonant tuning cavity, and the surface area of the cavity is chosen so the power density is 60 W/in^2. At this power density, heat transfer is by nucleate boiling, as shown in Figure 6.3. The surface temperature of the cavity is approximately 110°C and the power density can vary by a factor of two times without changing the cavity temperature by more than 2°C.

The vapor resulting from the boiling process rises from the bath, through the duct to the air cooled condenser. The vapor is condensed to liquid in the condenser and then returns through the duct to the liquid bath.

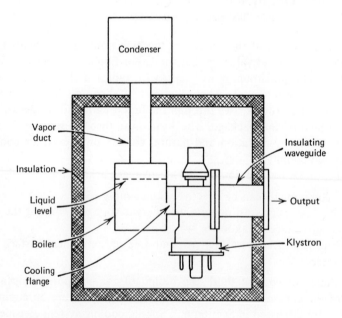

FIGURE 6.5
Evaporation cooling used to provide a constant temperature bath for a klystron oscillator

The boiling temperature of the water depends on the vapor pressure in the system, and therefore pressure must be held constant. This is achieved by the use of an expansion chamber on the condenser.

Expendable Cooling

Figure 6.6 shows a cutaway view of an aircraft electronics system mounted in a pod. The several electronics packages which make up the system are mounted on a rectangular water tank which runs down the length of the pod. The individual packages each transfer their heat by conduction to the water tank. The power density at the package to water tank interface is approximately 50 W/in². At this power density, as was shown in Figure 6.3, heat is transferred into the water by nucleate boiling.

Up to this point, this expendable system appears no different than the other applications of evaporation cooling discussed previously. However, the steam resulting from the boiling process is not condensed in this expendable system, but is simply exhausted to the surroundings. Large amounts of power can be dissipated with this simple expendable system. This approach avoids the difficult problem of providing an air cooled condenser which must work at high altitudes where air density is low. The operation time of the

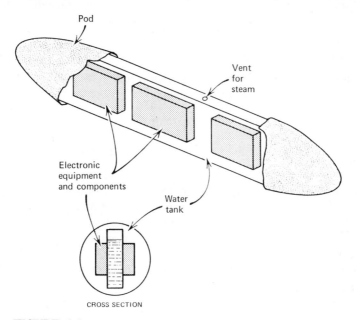

FIGURE 6.6
A pod mounted expendable cooling system for airborne electronics equipment

expendable evaporation cooling system is limited by the amount of water that can be carried in the tank, but this time can usually be made equal to the flight time of the aircraft, which is limited by the aircraft's fuel capacity. For the next flight, the water tank must be refilled.

Formulas for calculating the amount of coolant required by expendable evaporation cooling systems as a function of operating time are presented in Section 6.5.

6.2 DESIGN OF EVAPORATION COOLING

In designing electronic equipment to be cooled by evaporation, the following factors must be considered:

1. Choice of the cooling fluid.
2. Design of the electronic component surface where the evaporation takes place.
3. Design of the condenser.
4. The effect of pressure.
5. Equipment orientation.

Design information for heat transfer by evaporation, including detailed consideration of the above factors, is presented in this section.

Choice of the Cooling Fluid

The selection of the best fluid for evaporation cooling requires consideration of the following properties of the coolant:

1. Heat transfer capability.
2. Insulating properties.
3. Compatibility of the fluid with other elements in the electronic system.

If the fluid need not be an electrical insulator, for example, as in the case of the evaporation cooled high power transmitter tube discussed in the first example of Section 6.1, water is the best fluid. Water has several times greater heat transfer capability than the best dielectric fluids. Water is compatible with almost all materials used in electronic equipment and is of course low in cost.

When the fluid must be an electrical insulator, as well as providing evaporation cooling, water is not suitable. An example of this requirement is the liquid filled electronic package described in the second example of Section 6.1, where many components operating at different voltages are all mounted close together in the same package and are all cooled by the same fluid. For this application a dielectric liquid must be used. Not all dielectric fluids are

TABLE 6.1
Properties of Coolants Used for Evaporative Cooling

Property	Units	Coolant					
		Water	FC-78	FC-75	FC-43	Freon E2	Freon 113
Boiling temperature (at 1 atm)	°C	100	50	102	174	105	48
Maximum heat transfer by boiling	Watts/inch^2	870	225	270	350	94	100
Heat of vaporization	Watt hour/pound	285	12	11	9	9	19
Volume expansion	inch^3/inch^3/°C	$.2 \times 10^{-3}$	1.6×10^{-3}	1.6×10^{-3}	1.5×10^{-3}	1.5×10^{-3}	1.6×10^{-3}
Dielectric strength	kV/inch	—	430	550	560	340	310
Density	pounds/foot^3	62.4	160	110	117	103	96

suitable for evaporation cooling. To be suitable, the fluid must have a low boiling temperature, a high heat of vaporization, and low viscosity. In addition to having good insulating properties, the fluid must also be self-healing, that is, if voltage breakdown should occur, the dielectric characteristics of the liquid should be such that no conductive or corrosive substances are formed by the passage of the electric arc.

The most commonly used dielectric liquids for evaporation cooling are fluorocarbons (like FC-78, FC-75, and F-43 manufactured by 3M) and chlorofluorocarbons (like Freon E2 and Freon 113 manufactured by DuPont).

The properties of these dielectric liquids that are important for evaporation cooling are compared in Table 6.1, along with water.

The temperature of the electronic component being cooled by evaporation is always a few degrees above the boiling temperature of the liquid. As shown in Table 6.1, different boiling temperatures and therefore different component temperatures can be obtained by the proper choice of liquid. It should also be noted that any desired boiling temperature from 50°C to 175°C can be obtained by mixing FC-78, FC-75, and FC-43 in the proper proportions.

Figure 6.7 shows the temperature rise of an electronic component as a function of the power density at the component surface for three different fluorochemical coolants: FC-78, FC-75, and FC-43. Note that the higher the boiling temperature of the liquid, the higher the component temperature. Note also that when the critical power density is reached the component temperature rises almost without limit.

At low power densities heat is transferred from the components by natural convection. Temperature can be calculated using formulas similar to those of Chapter 3 for natural convection cooling with air. However, a detailed calculation of the liquid convection case is usually not necessary. The components will be cooled by liquid convection if the power density at their surface is low enough. If it is not, then boiling will occur and the component temperature will be from 5°C to 25°C above the boiling temperature of the liquid. As Figure 6.7 shows, evaporation cooling provides a order of magnitude increase in the amount of heat that can be transferred from the component, as compared to liquid convection cooling.

The temperature of an electronics component immersed in silicone oil is also shown in Figure 6.7. In the convection cooling range the cooling capability of the silicone oil and the fluorochemicals are approximately the same. However, the boiling temperature of the silicone oil is so high that evaporation cooling cannot occur below the safe operating temperature of most components. For this reason silicone oil is not a suitable liquid for evaporation cooling.

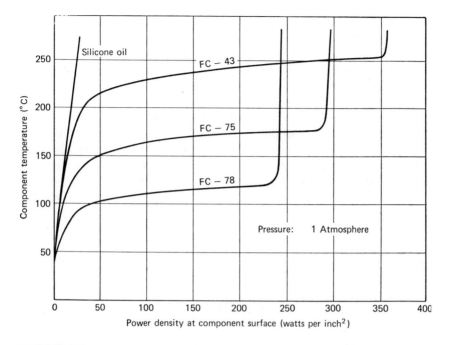

FIGURE 6.7
Component temperature as a function of power density for evaporation cooling with fluorochemical liquids

The volume expansion with temperature, the dielectric strength, and the density of the dielectric liquid must all be considered when designing a liquid filled electronics package. The significance of these properties, which are tabulated in Table 6.1, is discussed in Section 6.4.

Design of the Component Surface for Evaporation Cooling

The values of maximum power density given in Table 6.1 for the various liquids used for evaporation cooling should be taken as approximate values only. The maximum heat that can be transferred also depends on:

1. Whether the surface is horizontal or vertical.
2. The material of which the surface is made.
3. Whether the surface is oxidized or not.

The nature of evaporation cooling is such that immediate destruction of the electronic component will occur if the maximum power density is exceeded, because when the maximum is exceeded, the temperature of the component rises by several hundred degrees. The power density at the component surface must therefore never be allowed to exceed the maximum value, and because this maximum value is dependent on the several factors listed above, a safety factor should always be included in the design.

One way of providing the necessary safety factor, so that a hot spot will not cause thermal runaway, is shown in the photograph of Figure 6.1. The component surface incorporates thick vertical fins. The area of the base of the fins is great enough so that at the maximum power rating of the component all of the power could be transferred from the base area alone. The tip of the fins is always at a lower temperature than the base, because a temperature difference is required to conduct heat from the base to the tip. When water is used as the evaporative cooling liquid, for example, the base of the fins may be at 125°C, and the tips at 110°C. As can be seen from Figure 6.3, an order of magnitude less power is transferred from the fin tips than from their base. If the power density should momentarily exceed the maximum value at the base of the fins, the temperature of the base will rise as the base becomes insulated with a layer of vapor. Fortunately, however, the temperature of the fin tips will rise also, but not above the safe operating temperature of 125°C and the power will now be safely transferred from the tips of the fins rather than from the base.

To further insure that the layer of vapor does not form to insulate the component surface, horizontal slots are shown cut through the vertical fins in Figure 6.1. These horizontal slots serve to break up any vapor layers that form. As a general rule, the more complicated the shape of the surface, the less likely the chance that the insulating vapor layer will form.

The complicated geometry of the cooling surface described in the previous paragraph need only be used when the maximum rate of heat transfer must be obtained. If the design value of heat transfer per unit area is reduced to one-half the maximum value, heat can be safely transferred by evaporation without any special design of the component surface.

Condenser Design

Except for the special case of expendable cooling, the vapor resulting from the evaporation cooling process must be condensed in the condenser section of the equipment and returned to the liquid bath to repeat the boiling process. The design of the condenser for evaporation cooling systems is described in this section.

The condenser must condense the vapor as fast as it is generated by the boiling process. The internal surface of the condenser must be cooler than the vapor so that the vapor can condense on it. The temperature difference between the surroundings and the vapor in the condenser consists of three parts:

1. The temperature difference between the vapor and the inner surface of the condenser.
2. The temperature difference required to conduct heat from the inner surface of the condenser to the external surface.
3. The temperature difference required to transfer heat from the external surface of the condenser to the surroundings.

The temperature difference between the inner and outer surface of the condenser can be calculated from the design information on heat conduction given in Chapter 2. The temperature difference between the external surface of the condenser and the surroundings depends on the method of heat transfer being used—radiation and natural convection, forced air cooling, or forced liquid cooling—and can be calculated from the design information in Chapters 3 through 5, respectively.

The temperature difference between the vapor and the condenser inner surface can be calculated from Figure 6.8. This figure shows the heat transferred per unit area to a clean vertical surface as a function of the temperature difference between the vapor and condenser surface. Curves are shown for water and for the fluorochemical liquids. If the surface is oxidized, or if the orientation of the condensing surface is not vertical, the temperature difference between the vapor and the condenser surface will be slightly different than the values shown in Figure 6.8. However, Figure 6.8 is adequate for most design estimates.

If water vapor is being condensed, and the area of the condenser is such that the power density is 40 W/in^2, then the temperature difference between the steam and the condenser surface is 4°C. If FC-75 vapor is being condensed at this same power density, the temperature difference between the vapor and the condenser walls is 63°C. For most designs, the inner surface of the condenser should be made large enough to limit the temperature difference between the vapor and surface to 10°C.

A simple example will illustrate the use of Figure 6.8. Assume that a high power transmitter tube which dissipates 1 kW of power is being cooled by evaporation of water at atmospheric pressure. If the anode area of this high power tube is 2 in^2, then the power density at the anode surface is 500 W/in^2, which is safely below the maximum power density for evaporation cooling. The outer surface of the condenser is cooled with forced air. The temperature difference between the cooling fins and the surroundings is 65°C. If the surroundings are at 20°C and a 5° temperature rise is required for conducting 1 kW of heat from the inner to the outer surface of the condenser, then the inner surface of the condenser will be at 90°C. The steam from the evaporation

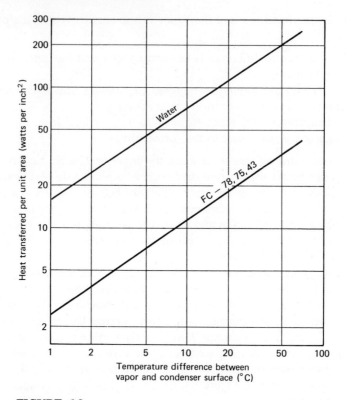

FIGURE 6.8

Heat transfer per unit area in a condenser as a function of the temperature difference between the vapor and the condenser walls

cooling process is at 100°C so the temperature difference between the steam and the condenser surface is 10°C. From Figure 6.8 the power that can be transferred by the condensing steam is 70 W/in². Therefore, the condenser surface need have an area of only 14 in² to provide enough surface to condense the steam as fast as it is generated.

In most evaporation cooling systems, the temperature of the liquid bath is at approximately the boiling temperature of the liquid. This is the case in the four typical designs discussed in Section 6.1 where the liquid is not in contact with the cooled condenser surface. However, in some liquid filled electronic packages, where heat is removed from the entire surface of the package, the liquid can be maintained below its boiling temperature. In this case the vapor from the boiling process is condensed in the liquid bath itself. This approach removes the limitation that the condenser surface can be only the top surface of the package.

The Effect of Pressure

All of the data on evaporation cooling presented thus far in this chapter has been for designs where the pressure in the system has been equal to 1 atm (14.7 psi). This pressure is easily achieved by venting the system. The temperature of the component as a function of power density when water is used as the cooling liquid as shown in Figure 6.3, and when fluorochemical liquids are used as coolants, as shown in Figure 6.7, were all for the case when the pressure in the evaporation cooling system was equal to 1 atm.

Unfortunately, it is not always possible to vent the evaporation cooling system to maintain atmospheric pressure, for example, if the equipment must be operated in an isolated location where servicing to replenish the liquid which evaporates is not possible, or if the equipment must be mounted in a sealed package. When the system is sealed, the internal pressure will vary depending on the power dissipated within the package, the external cooling conditions, and the temperature of the surroundings. The effect of pressure on evaporation cooling is discussed in this section.

The most important effect of pressure in evaporation cooling systems is that the pressure controls the boiling temperature of the liquid. Figure 6.9 shows the boiling temperature of water and the fluorochemical and Freon liquids as a function of pressure. With evaporation cooling, the temperature of the electronic component is always a few degrees above the boiling temperature of the coolant. Therefore, the component temperature, because it is directly related to the boiling point of the liquid, is controlled by the system pressure.

The effects of pressure in a sealed evaporation cooling system can best be illustrated by an example. Assume the system uses FC-75 as a coolant. Before sealing, the equipment is operated at full power. The package is completely filled with the liquid and its saturated vapor. Since the package is still unsealed the pressure is 1 atm. As shown in Table 6.1, the boiling temperature of the FC-75 is 102°C at 1 atm. The temperature of the condenser surface must be lower than the boiling point of the liquid so that the vapor can condense as fast as it is formed. In this example, the condenser area will be chosen so that a 10°C temperature difference is required between the temperature of the vapor and the temperature of the condenser walls. As was shown in Figure 6.8, at this temperature difference approximately 11 W/in^2 can be transferred from the vapor to the condenser surface. Since the boiling temperature at atmospheric pressure of FC-75 is 102°C, the inner surface of the condenser must operate at 92°C and assuming a 2°C difference between the inner surface of the condenser and its external cooling fins, the cooling fin temperature is 90°C.

While the system is operating under the above conditions, it is sealed. As

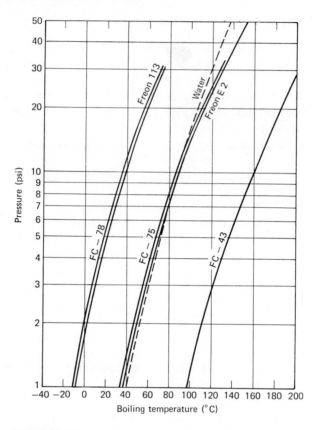

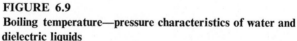

Boiling temperature (°C)

FIGURE 6.9
**Boiling temperature—pressure characteristics of water and
dielectric liquids**

long as the power being transferred from the components remains unchanged
and the condenser temperature remains unchanged, the pressure inside the
container remains at one atmosphere and the boiling temperature remains at
102°C.

If, after the system has been sealed, the power dissipated by the electronic
components changes, or if the temperature of the condenser changes, the
pressure and the boiling temperature of the liquid will both change. Figure
6.10 shows how the internal pressure and the boiling temperature vary with
these parameters. Curve 1 shows the internal pressure and the boiling
temperature of the FC-75 liquid as a function of condenser temperature for
the case when the power dissipated by the components remains constant.
The temperature of the condenser depends both on the temperature of the
surroundings and on the amount of external cooling applied. For example,

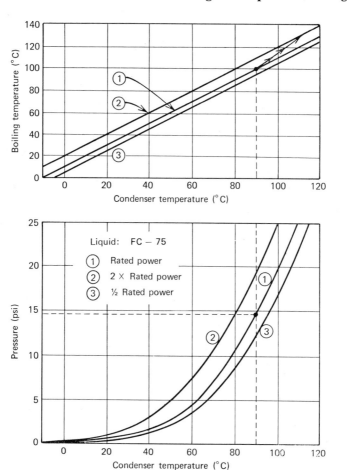

FIGURE 6.10
Boiling temperature and pressure as a function of condenser temperature for a sealed liquid filled electronics package

if the condenser is cooled on its external surfaces by forced air, its temperature will depend on the amount of air flow. Since the power that is transferred from the vapor to the condenser is constant, the temperature difference between the condenser and the vapor remains constant. Therefore, the boiling temperature of the liquid will vary directly as the condenser temperature, and this is shown in curve 1. The pressure inside the system correspondingly adjusts as required so that the boiling point of the liquid always remains 10°C below the condenser temperature. As the condenser temperature is lowered, for example, by applying additional cooling, the boiling temperature decreases and the pressure inside the system decreases rapidly. Since the component part temperature is directly proportional to the boiling temperature of the liquid, the component part temperatures decrease also. As the condenser temperature is raised above the value at which the system was sealed, the boiling point of the liquid rises above its atmospheric pressure value in order to maintain the constant temperature difference between the vapor and the condenser surface, and consequently the pressure inside the system also rises. This fact shows the importance of operating the system at the maximum condenser temperature at the time of sealing to prevent the pressure from rising to extremely high values and rupturing the container.

Curve 2 of the Figure 6.10 shows a case where the power dissipation is increased above the value at which the system was sealed. In order to transfer the additional power from the vapor to the condenser, the temperature difference between the vapor and the condenser must increase and so the boiling temperature of the liquid must increase. The pressure inside the system adjusts appropriately so that the liquid will now boil at the required higher temperature.

In a typical system, when the power being dissipated by the components increases, the condenser temperature probably has to increase also to transfer this additional power to the surroundings. Consequently, the operating point moves as shown by the arrow from curve 1 to curve 2 and a large increase in pressure occurs. This again emphasizes the importance of operating the system at the maximum power level before sealing.

Curve 3 shows the effect if the power dissipation decreases from the value when the system was sealed. In this case the temperature difference between the vapor and the condenser walls must be less, so the internal pressure adjusts to lower the boiling temperature of the liquid.

Although the major effect of pressure in evaporation cooling systems is to change the boiling temperature of the liquid, one other important effect must also be considered. Pressure controls the maximum power per unit area that can be transferred. This effect is illustrated in Figure 6.11, where the maximum power density that can be transferred by evaporation cooling is shown as a function of pressure when water is used as the evaporating liquid.

Similar curves are obtained for liquids other than water.

When the pressure in the evaporation cooling system is 1 atm, the maximum power density that can be achieved by evaporation cooling, before the vapor films begins to form around the component, is 870 W/in^2. This is the value shown in Figure 6.3 and Table 6.1. As the pressure in the evaporation cooling system is increased, the maximum power density that can be achieved increases as shown in Figure 6.11. For example, when the pressure in the evaporation cooling system increases to 100 psi, the maximum power density approximately doubles. Actually, it is not practical to take advantage of this

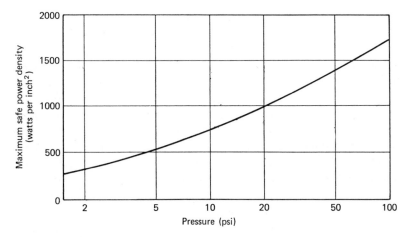

FIGURE 6.11
Maximum safe power density for evaporation cooling with water as a function of pressure

higher power density by increasing the pressure in an evaporation cooling system, because this requires that the electronic components must be designed to withstand high pressure and the system enclosure must be a high pressure container.

The more important portion of the curve is the pressure region below atmospheric, where, as the pressure decreases, the maximum power that can be transferred by evaporation cooling decreases. For example, when the pressure in the system decreases to 2 psi, the maximum safe power density decreases to approximately one-third of its atmospheric pressure value, that is to 300 W/in^2. As was shown in Figure 6.10, the pressure in a sealed evaporation cooling system decreases rapidly when the condenser temperature is decreased below the value at which the system was sealed. This decrease in pressure is accompanied by a decrease in boiling temperature and consequently a decrease in component temperature which is desirable. However, as shown in Figure 6.11, the maximum power density that can be safely transferred from the component decreases also, and this fact must be included in the design. The power density at the component surfaces must be less than the safe maximum, not only at atmospheric pressure where the system is sealed, but also at the lowest pressure at which the system will operate.

Equipment Orientation

Regardless of the type of evaporation cooling system, whether it is vented to the atmosphere or sealed, whether it is used to cool a single high power component or a complete equipment in a liquid filled electronics package, it can operate in *one* orientation only, which is with the condenser at the highest point in the system. The basic principle of evaporation cooling is that heat is transferred from the electronic components by evaporating the liquid and the resulting vapor rises and is condensed in a condenser. The condenser must therefore be at the highest point in the equipment or the vapor cannot reach it.

This fact, that electronic components must be at a lower point in the system than the condenser, limits the use of evaporation cooling to installations where the orientation of the equipment can always be maintained. Even if the evaporation cooling system is sealed, it still must be operated with the condenser on the top surface of the package. If an electronic equipment must be operated in any orientation, then the simple evaporation cooling methods discussed in this chapter cannot be used. However, a modified form of evaporation cooling, using heat pipes, will permit operation in any orientation. The design and use of heat pipes for cooling electronic equipment is discussed in detail in Chapter 7.

6.3 COOLING SYSTEM CONSIDERATIONS FOR HIGH POWER COMPONENTS

A simplified drawing of an evaporation cooling system for high power electronic tubes was shown in Figure 6.2. A more complete drawing of a practical evaporative cooling system, including all the necessary components, is shown in Figure 6.12. The elements of the complete evaporation cooling system are:

- Boiler.
- Insulating tubing.
- Control box.
- Equalizer line.
- Condenser.
- Piping.
- Pressure interlock.

The boiler supports the power tube and contains the water used for cooling. The tube's anode flange must seal securely against the top of the boiler so that the steam is exhausted through the steam outlet. In most cases the anode of the tube, and consequently the boiler, is operated at a high voltage relative to ground; therefore, the boiler must be electrically insulated from the rest of the system. To accomplish this, the boiler is mounted on insulators and all steam and water connections to the boiler are made through pyrex insulating tubing. The length of the insulating tubing is dependent upon the voltage that is applied, the purity of the water, and the volume of the returned cooling water. A 2 ft section of pyrex tubing can prevent voltage breakdown up to 20 kV and will have negligible leakage current.

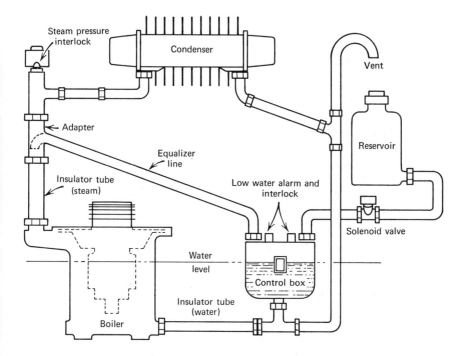

FIGURE 6.12
Schematic drawing of a complete evaporation cooling system for high power transmitter tubes

The function of the control box is to monitor and control the water level in the boiler and to serve as a partial reservoir. The control box is an airtight vessel containing an overflow syphon and two float switches. When the water level drops approximately 1/4 in. below the recommended level, the first float switch is actuated and turns on a warning alarm and a solenoid controlled water valve to admit more water from the reservoir. The second float switch is operated if the water should drop 1/2 in. below the optimum level and is used to remove the power to protect the tube from damage.

The control box must operate at the same pressure as exists in the boiler. To accomplish this, the system should be fitted with an equalizer line, which is a length of tubing which connects the steam side of the system with the top of the control box.

The condenser may be air cooled or liquid cooled. It may be mounted in any position that will allow the condensed water to flow freely by gravity to the boiler return line, and must of course be mounted above the level of the boiler. The piping for the steam or the return water should be of copper or brass. The size depends on the power level and the volume of generated steam. For example, for a tube dissipating 8 kW of power, the piping inner diameter should be 1 3/4 in. The piping is insulated from the boilers which are at high voltage by the pyrex tubing.

The system should be provided with a pressure interlock on the steam side of the condenser. This interlock is set about 1/2 psi above atmospheric pressure, and is used as a power interlock that will turn off the system if any abnormal steam pressure develops due to constrictions in the condenser or piping.

All of the necessary parts of the evaporation cooling system described above are simple low cost plumbing fixtures. They include no moving parts like pumps to consume power or wear out, and the entire system operates at atmospheric pressure. The design of the electronic component can be simplified and its size reduced by the high heat transfer rates made possible by evaporation cooling. The high power electronic component, which is the most expensive part of the complete equipment, can therefore be reduced in cost. It is for these economy reasons that evaporation cooling has found widespread usage in high power radio transmitter equipment.

Evaporation cooling is effectively used on high power transmitter tubes with plate dissipations up to several hundred kilowatts. Figure 6.13 shows a family of vapor cooled high power tubes with plate dissipations from 8 kW to 250 kW.

FIGURE 6.13
A family of evaporation cooled high power transmitter tubes (Photo courtesy of Varian Associates)

6.4   LIQUID FILLED ELECTRONICS PACKAGES

Liquid filled electronics packages offer the following advantages:

1. All of the components in an electronics equipment can be simply cooled by immersing the entire equipment in the liquid. The problems of thermal interfaces and conduction of heat from the components to the package are avoided.
2. The electronic components can operate at different voltages without danger of high voltage breakdown if a dielectric oil is used as the coolant.
3. Very effective cooling can be obtained with high rates of heat transfer per unit area.

To obtain full advantage of the simplicity, voltage insulation, and high heat transfer rate offered by evaporation cooling in liquid filled electronics packages, the following design factors must be considered:

1. Choice of cooling liquid.
2. Heat transfer from the components.
3. Condenser design.
4. Expansion of the liquid with changes in external temperature.
5. Internal pressure.
6. Minimizing the volume and weight of the liquid.
7. Sealing procedures.

Typical designs of liquid filled electronics packages, which consider the above factors in various ways, are shown in Figures 6.14A through F. The packages shown in Figures 6.14A through C are only partially filled with liquid to permit thermal expansion of the liquid as the temperature of the surroundings changes. The packages shown in Figures 6.14D through F are completely filled with liquid, and the expansion of the liquid with temperature is compensated for by equipping the packages with a flexible wall or an expansion bellows. Package designs A, B, D, and E use the inside top surface of the package as the condenser where the vapor resulting from evaporation at the components is condensed. The heat is then transferred from the package top to the surroundings by forced air cooling in designs A and D and by forced liquid cooling in designs B and E. In design C the package is connected to an external forced air cooled condenser, and the vapor resulting from the boiling process at all of the components in the oil bath rises to the condenser, and after condensing returns to the package. In design F, where the package is completely filled with liquid, the liquid is maintained below its boiling temperature by the forced liquid cooling coil. In this case the vapor condenses in the liquid itself rather than on the top surface of the package.

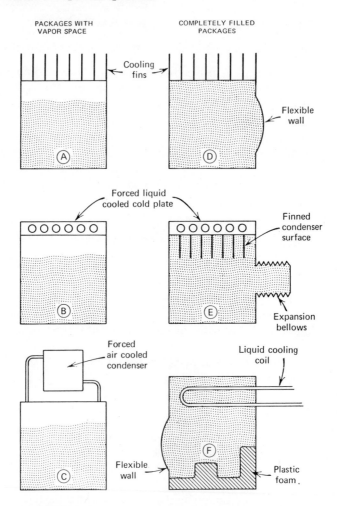

FIGURE 6.14
Liquid filled electronics packages

Design E illustrates the use of fins on the inner condenser surface to increase the area for condensing the vapor. Design F illustrates the use of a plastic foam filler inside the package to minimize the volume and weight of the dielectric liquid. Further details on these package designs are discussed in subsequent paragraphs.

The coolant liquid must have good dielectric properties and must have a boiling temperature lower than the safe operating temperature of the components to be cooled. Suitable dielectric liquids for evaporation cooling were discussed in Section 6.2, and the important thermal properties of commonly used liquids were tabulated in Table 6.1. The major difference between the dielectric liquids shown in Table 6.1 is their boiling points. The higher the boiling point of the liquid, the higher the condenser temperature can be and the more effective the condenser can be in transferring heat from its external surface to the surroundings. On the other hand, the lower the boiling point of the liquid, the lower will be the component temperature, since the temperature of the components must always be a few degrees above the boiling point of the liquid.

The heat transfer from the individual components can be calculated from the design information presented in Section 6.2. For example, consider a transistor which has a case area of .5 in^2, in a package filled with FC-75 liquid. If the transistor dissipates 10 W of power, then the power density at the transistor surface is 20 W/in^2. From Figure 6.7, which shows the temperature of the electronic components as a function of the power density at their surface when they are immersed in the three fluorochemical liquids, the transistor temperature at this power density would be 130°C. At this power density the heat is being transferred by natural convection. If the transistor heat dissipation were increased to 25 W, then the power density would be 50 W/in^2. Again using Figure 6.7, the component temperature would be 150°C and the heat transfer would be by evaporation. The transistor case temperature could be lowered from 150°C to 105°C by using FC-78, which has a lower boiling temperature than FC-75.

The location of the heat generating elements in each electronic component must be carefully considered in determining the effective area of the component surface when the power density is being calculated. Some areas of the component, as for example the top of a transistor case, may not be in good thermal contact with the heat generating elements of the component and so these areas cannot be included in the calculation of effective surface area. The part of the component where the heat is generated may need to be finned, as discussed in Section 6.2, to provide a margin of safety for transferring heat by evaporation if the power density is near the maximum safe limit. In the example described above, if FC-75 is used as the liquid filling the package, the maximum safe power density at the component surface is approximately 290 W/in^2 at a pressure of 1 atm, as shown in Figure 6.7. The effect of system pressure on the maximum safe power density must also be considered. As was shown in Figure 6.11, the maximum safe operating power density decreases by a factor of about three when the pressure is lowered from atmospheric to a few pounds per inch^2. Since the pressure usually always decreases in a liquid filled electronics package as the temperature of the surroundings change, the maximum safe power density must be reduced accordingly. For the case of FC-75, the power density at the component surface should probably not exceed 100 W/in^2 in a sealed electronics package. The power density of the transistor surface, even when it is operating at the 25 W level, is only 50 W/in^2 and well below this safe limit so no special finned surfaces are necessary.

The design of the condenser of a liquid filled electronics package has two parts:

1. Design of the inner surface of the condenser (which is usually the inside top of the package) to condense the vapor as fast as it is formed.
2. Design of the external surface of the condenser to transfer heat from the package to the surroundings.

The second part of the design problem, the design of the external surface of the condenser, is usually the most difficult. Heat can be transferred from the electronic package by radiation and natural convection, forced air cooling, or forced liquid cooling. The proper design using any one of these methods can be obtained from the information presented in Chapters 3, 4, and 5 respectively. Uses of these various types of cooling of the external surface of the package are illustrated in Figures 6.14A, B, D, and E. In designs A and D the outside of the top surface of the package is finned so that the heat can be transferred to the surroundings by radiation and natural convection or by forced air cooling. In designs B and E the top surface of the package is cooled with forced liquid. The design approach of using a liquid filled electronics package where the heat is transferred from the package by forced liquid cooling is a much simpler way of cooling the many components that comprise the equipment, rather than circulating the liquid through each of the components individually.

In some cases, the total power dissipated in the liquid filled electronics package may be greater than can be easily removed by circulating cooling air over the top of the package. In this case, an external forced air cooled condenser can be used as illustrated in the design shown in Figure 6.14C. The design of this condenser can be optimized for cooling, since there is no requirement that it fit on the top of the electronics package.

The amount of inner surface area required to condense the vapor resulting from the boiling process can be determined from Figure 6.8, which shows the power that is transferred per unit area of condenser surface as a function of the temperature difference between the vapor and the condenser surface. For example, if the surface area of the top of the electronics package is 25 in^2 and the total power dissipated in the equipment is 250 W, then the power density at the condenser surface is 10 W/in^2. If a fluorochemical liquid is used in the package, the temperature difference between the vapor and the condenser surface is 8°C, as shown in Figure 6.8.

If the power density at the inner surface of the condenser is so high that the temperature difference required to transfer the power exceeds 10°C, a better design would be to fin the inner surface of the condenser as shown in Figure 6.14E and thereby reduce the power density.

In all the designs of liquid filled electronics packages except that shown in Figure 6.14F, care must be taken to insure that no air is mixed with the vapor in the package. The presence of air significantly reduces the amount of heat that can be transferred as the vapor condenses, since the air forms an insulating layer over the condenser surface. In the design shown in Figure 6.14F, the liquid cooling coil maintains the liquid in the package below its boiling temperature. In this case, the liquid boils at the surface of the components and the vapor then condenses in the bulk of the liquid. When this approach is used, the area of the top surface of the electronics package is no longer critical, because it does not control the condensation process, and the presence of air mixed with the vapor in the package is not critical.

The change in volume of the dielectric liquids with temperature is very large. As shown in Table 6.1, the rate of change of volume with temperature is approximately eight times greater for the dielectric liquids than for water.

Many electronic equipments, especially those designed for military use, must operate over a wide temperature range, for example from $-55°C$ to 90°C. The change in volume of a dielectric liquid over this temperature range is $1.6 \times 10^{-3}/°C \times 145°C = 23\%$. If the electronics package is completely filled with liquid when the liquid is at its maximum temperature, then the liquid level will decrease by 23% at the minimum operating temperature. The electronic components must remain immersed in the fluid at all times to prevent voltage breakdown and to provide adequate heat transfer. If the electronic package is designed with the liquid filling only part of the package at low temperatures, then the components must be located in the lower three-quarters of the package so that the liquid level will at no time drop below the components. An alternate approach is to design the package with a flexible wall or an expansion bellows, as is shown in Figures 6.14D, E, or F. As the liquid contracts when the equipment is operated at low temperature, the flexible wall or the bellows contracts so that the package will be always filled with liquid.

As described in Section 6.2, the pressure inside a sealed package depends on the power being dissipated by the electronic components, the amount of cooling applied to the condenser, and the temperature of the surroundings. If the dissipated power decreases from the value at which the package was sealed, or if the condenser temperature decreases because of improved cooling or reduced temperature of the surroundings, then the pressure inside the electronics package may decrease to only a fraction of a pound per inch^2. Therefore, all liquid filled electronics packages must be designed to withstand an external pressure of 1 atm.

If the dissipated power increases above the value at which the package was sealed, or if the condenser temperature increases due to an increase in the temperature of the surroundings or reduced cooling on the external surface of the package, the pressure inside the package will rapidly rise above 1 atm as was shown in Figure 6.10. Therefore, the liquid filled electronics package must be equipped with a pressure release valve, which will open when the pressure exceeds a safe limit of approximately 2 atm, to prevent damage to the electronic components in the package or rupture of the package itself.

The dielectric liquids used for liquid filled electronics packages are both expensive and heavy. As shown in Table 6.1, the density of the dielectric liquids is approximately two times as great as water. Therefore, the volume of liquid used in the package should be minimized to save on overall equipment cost and to reduce the weight of the package. A practical means of reducing the volume of the required liquid is shown in Figure 6.14F. The void spaces in the package where components cannot be practically located are filled with lightweight plastic foam. The dielectric liquid still surrounds all the components but the foam replaces the heavy liquid in all spaces where the liquid is not needed for cooling or dielectric insulation purposes.

Because air, water, or other impurities in the dielectric liquid will degrade the performance of the condenser in a liquid filled electronics package and may cause voltage breakdown problems, the package must be carefully filled so that no impurities are present in the liquid. The first step in this filling procedure is to evacuate the container with the components inside to remove all air, water, and other vapor impurities. Then the package is filled with the dielectric liquid. The equipment is then operated at the maximum power level and the maximum condenser temperature that the system will see in actual use. This maximum condenser temperature is obtained by operating the equipment with the least external cooling and with the highest temperature of the surroundings. At this level of operation, the liquid level is adjusted to fill the entire package and the equipment is then sealed. By using this filling procedure, the pressure inside the package will always be at 1 atm or less and all noncondensable impurities will be removed.

6.5   EXPENDABLE COOLANT SYSTEMS

Considerable weight, volume and cost savings can be realized by using evaporation cooling with an expendable coolant. In this design, the heat is transferred from the electronic equipment by evaporation and the vapor is simply exhausted to the surroundings. This simple cooling system eliminates the need for cooling fans or for liquid pumps and air-to-liquid heat exchangers. The expendable coolant approach is particularly useful in airborne electronic equipment where normal cooling methods are difficult to apply because of the poor cooling capability of high altitude air and where the equipment operating time is limited by the flight time of the aircraft.

Although any cooling liquid listed in Table 6.1 could be used in an expendable coolant system, water is the only practical liquid to use. Water is considerably less expensive than the other liquids. It has a much higher heat of vaporization and a lower density, so less volume and weight of water are required to dissipate a given amount of power than with any other fluid. If the equipment to be cooled by an expendable cooling system consists of a number of components which must be operated at different voltages, the equipment should be designed as a liquid filled electronics package as discussed in Section 6.4. The complete package can then be cooled by an expendable coolant system using water as the expendable liquid in the secondary system. This approach was illustrated in Figure 6.6, where the individual sealed packages of an expendable cooling system were shown mounted on a water tank in an aircraft pod.

The weight and/or volume of water required by an expendable coolant system depends on the power that must be dissipated by the system and the required operation time. The necessary weight and volume of the water are given by Equation 6.1:

$$\text{Weight of water (pounds)} = 5.9 \times 10^{-5} Q\gamma$$
$$\text{Volume of the water (inch}^3) = 1.6 \times 10^{-3} Q\gamma \tag{6.1}$$

where:

Q is the power that must be dissipated (watts).
γ is the system operating time (minutes).

For an operating time of 1 hr (60 min), Equation 6.1 shows that approximately 3.5 lb of water are required per kilowatt of dissipated power. This weight of water occupies a volume of 96 in^3, which for example, would require a tank only 2.5 in. wide $\times$ 3 in. high $\times$ 13 in. long. The size and weight of this water filled tank are considerably less than would be required by a conventional forced air or forced liquid cooling system.

A related type of expendable cooling system can be designed using the heat of fusion of solid coolants rather than the heat of vaporization of water. These heat of fusion systems provide less cooling time per pound or volume of coolant, but they eliminate the problem of venting the vapor to the atmosphere. In these systems the coolant tank is filled with a solid, such as paraffin wax or an inorganic salt. The heat from the electronic equipment melts the solid at constant temperature. When all the solid has been melted the equipment must be turned off, but after a sufficient cooling time has elapsed to allow the material to solidify, the equipment can be operated again.

Table 6.2 compares the heat sink temperature and the weight and volume of coolant material required per kilowatt-hour for evaporation cooling with water and for two heat of fusion coolant systems.

TABLE 6.2
Expendable Coolant Systems

Type of System	Coolant	Heat Sink Temp. (°C)	Weight Per Kilowatt-Hour (pounds)	Volume Per Kilowatt-Hour (inch³)
Evaporation	Water	100	3.5	96
Heat of fusion	Paraffin	52	60	1800
Heat of fusion	Inorganic salt	90	50	700

6.6 EXAMPLES

Evaporation cooling of electronic equipment can be designed from the equations, figures, and tables of this chapter. The correct use of this design information is illustrated in this section by the following examples:

- An evaporation cooled high power transmitter tube.
- A liquid filled electronics package.
- An airborne electronic equipment using an expendable coolant.

EXAMPLE 1
An Evaporation Cooled High Power Transmitter Tube

A family of high power transmitter tubes which are designed for cooling by evaporation was shown in Figure 6.13. The complete evaporation cooling system for any of these tubes was shown in Figure 6.12. The lowest power member of this family, which is shown on the right-hand side of the figure, is capable of dissipating 8000 W from its anode by evaporation cooling. The cooling design of this tube will be calculated in this example.

The first step is to determine the external dimensions of the tube anode where the evaporation process occurs. A sketch of the preferred anode geometry is shown in Figure 6.15. As discussed in Section 6.2, this special geometry is necessary for the cooling of high power components by liquid

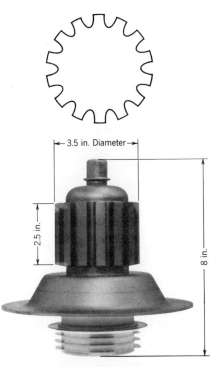

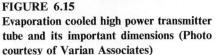

FIGURE 6.15
Evaporation cooled high power transmitter
tube and its important dimensions (Photo
courtesy of Varian Associates)

evaporation to insure that a momentary increase in power density will not allow the surface to be insulated by a layer of vapor and cause immediate burnout. The preferred design is to use thick fins on the component surface and make the area at the base of the fins large enough to transfer all the power. If the power density should momentarily increase, the base of the fins will be covered by an insulating layer of vapor and the temperature will begin to rise, but the point where heat is transferred will move out to the tip of the fins and the thermal runaway will not occur.

If water is used as a coolant, the maximum power that can be transferred by evaporation, at a pressure of 1 atm, is 870 W/in² as shown in Table 6.1. The area at the base of the fins must be 8000 W divided by 870 W/in² or 9.2 in². The anode length is 2.5 in. so the anode diameter at the base of the fins should be

$$2 \times \frac{9.2 \text{ in}^2}{\pi \times 2.5 \text{ in.}} = 2.35 \text{ in.}$$

The factor of two is used because only half of the circumference is available at the base of the fins. The area of the tip of the fins is approximately as great as the area at the base, so all the power could also be transferred from the tips of the fins.

The next step is to calculate the surface area of the inside of the condenser. Assuming that the cooling of the external surface of the condenser has been designed so that the inner surface can be maintained at 90°C, the temperature difference between the steam and the condenser walls is 10°C. From Figure 6.8, 70 W of power can be transferred per inch^2 of surface. The inner surface area of the condenser must, therefore, be

$$\frac{8000 \text{ W}}{70 \dfrac{\text{W}}{\text{in}^2}} = 115 \text{ in}^2.$$

The high power transmitter tube shown in Figure 6.15 was designed for evaporation cooling with water. If the tube were to be evaporation cooled with a fluorochemical liquid, the anode dimensions would have to be increased. As shown in Table 6.1, the maximum heat that can be transferred by evaporation cooling with FC-75 is 270 W/in^2. Therefore, the anode area would have to be increased by approximately three times.

EXAMPLE 2
A Liquid Filled Electronics Package

Figure 6.16 shows a package of card-mounted electronic components cooled by evaporation. One end of the package has been removed to show the internal details. Cooling of a similar rack of card-mounted components by natural convection and by forced air was discussed in Example 3 of Chapter 3 and Example 3 of Chapter 4, respectively. In this present example, evaporation cooling of the same configuration will be calculated.

The equipment shown in Figure 6.16 consists of four 4 in. × 6 in. × 1/16 in. thick circuit boards. The electronic components are mounted on one side of each of the boards. The four cards are mounted in a sealed container 5 in. wide × 8 in. long × 5 in. high.

The factors that must be considered in the evaporation cooling design of this equipment are:

1. Choice of the cooling liquid.
2. Heat transfer from the components.
3. Condenser design.
4. Internal pressure.
5. Expansion of the liquid with changes in external temperature.
6. Weight of the liquid.

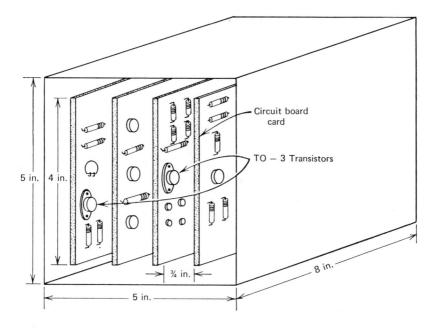

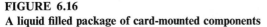

FIGURE 6.16
A liquid filled package of card-mounted components

The first step is to select the dielectric liquid. In order to minimize component temperature the fluorochemical with the lowest boiling temperature, FC-78, will be chosen. From Table 6.1, the boiling temperature of FC-78 at a pressure of 1 atm is 50°C.

The next step is to calculate the temperature of the components. The major heat generating components are two 20 W TO-3 transistors, one of which is mounted on each of two of the circuit boards. The effective area of the TO-3 transistor case is only .5 in² which is the area of its mounting flange. The total surface area of the transistor is greater, but the rest of the area is not in good thermal contact with the transistor chip. The power density at the mounting surface of each transistor is therefore 40 W/in². From Figure 6.7, transfer of heat is by evaporation and the temperature of the transistor cases is 100°C.

The next step is to calculate the temperature of the inner surface of the top of the package which serves as the condenser surface. The total power to be transferred is 40 W and the area is 5 in. × 8 in., so the power density at the condenser surface is 1.0 W/in². From Figure 6.8, the temperature difference between the vapor and the package wall is less than 1°C, so the package wall is at approximately 50°C.

Heat transfer from the outer surface of the package must be calculated next. If the walls of the package are made thick enough (for example, 1/4 in. if the package is aluminum) the heat can be conducted to all the walls of the package with a temperature difference of less than 10°C. Then all the walls can serve as the outer area of the condenser for transferring heat to the surroundings, although only the top of the package can serve as the inner area of the condenser.

In Example 4 of Chapter 3, the amount of heat that could be transferred by radiation and natural convection from an electronics package was calculated. The package dimensions were the same as in this example, and if the outer surface of the package is at 50°C, and the surroundings are at 20°C, approximately 40 W of power can be transferred up to an altitude of 30,000 ft. Therefore, in this example of a liquid filled package, the external surface area of the package is large enough so all the heat generated in the package can be transferred to the surroundings by radiation and natural convection.

The next factor to consider is the internal pressure in the package. The package will be sealed under full power operating conditions with its outer surface at 50°C. The pressure inside the sealed package will still be 1 atm as long as these conditions prevail. The boiling point of the liquid will be 50°C and as calculated above, the temperature of the transistor cases will be 100°C. If the temperature of the outer walls of the package is reduced, as for example when the package is operated at sea level instead of at an altitude of 30,000 ft, the pressure inside the package will decrease. If, for example the temperature of the package walls is lowered by 15°C, the temperature of the transistors will be lowered by 15°C also. From Figure 6.9, the pressure inside the package will be lowered to about 8 psi.

The maximum power per unit area that can be transferred by evaporation will decrease, but because the transistors are operating well below the maximum safe power density, this effect of pressure will cause no problem.

The next factor to consider is the change in volume of the dielectric liquid with temperature. If liquid completely fills the package when it is sealed at 50°C, and if the equipment is subsequently operated with the package at 0°C, the liquid will contract. From Table 6.1, the change in volume is $1.6 \times 10^{-3}/°C$, so at 0°C the liquid has contracted 8%. The liquid level will be .08 × 5 in. or .4 in. from the top of the package at this low temperature, and the circuit boards must be mounted below this level so that they will remain covered with dielectric liquid at all times.

The final step is to calculate the weight of the dielectric liquid, which constitutes a major part of the weight of this electronics package. From Table 6.1, the density of FC-78 is 160 lb/ft³. Assuming the liquid occupies 90% of the package volume, the weight of the liquid is:

$$\frac{.9 \times 5 \text{ in.} \times 5 \text{ in.} \times 8 \text{ in.}}{(12)^3 \dfrac{\text{in}^3}{\text{ft}^3}} \times 160 \frac{\text{lb}}{\text{ft}^3}$$

$$= 16.5 \text{ lb.}$$

EXAMPLE 3
An Airborne Electronic Equipment Using an Expendable Coolant

A drawing of an aircraft electronics system using expendable cooling was shown in Figure 6.6. The several electronic packages that make up the equipment are mounted on a rectangular water tank which extends down the length of the aircraft pod. The power density at the package to water interface is 50 W/in^2 and heat is transferred by boiling the water. The vapor is expended to the surroundings through the vent hole.

If the total power dissipation of the equipment is 5000 W, and the mission time is 2 hr, the required weight and volume of water can be calculated from Equation 6.1.

$$\text{Weight of water} = 5.9 \times 10^{-5} Q\gamma$$

$$= 5.9 \times 10^{-5} \times 5000 \text{ W} \times 120 \text{ min}$$

$$= 35 \text{ lb}$$

$$\text{Volume of water} = 1.6 \times 10^{-3} Q\gamma$$

$$= 1.6 \times 10^{-3} \times 5000 \text{ W} \times 120 \text{ min}$$

$$= 960 \text{ in}^3$$

where:

Q is the power that must be dissipated and is 5000 W.
γ is the system operating time and is 120 min.

6.7 REFERENCES

Useful references for evaporation cooling electronic equipment are as follows:

1. McAdams, W. H., *Heat Transmission*, McGraw Hill, New York, 1954, Chapter 13, pp. 325–367 and Chapter 14, pp. 368–409.

2. Armstrong, R. J., *Evaporative Cooling with Freon Dielectric Liquids*, Dupont Applications Report EL-11, Wilmington, Del., 1966.

3. Sutherland, R. I., *Care and Feeding of Power Grid Tubes*, Eimac Division of Varian, San Carlos, Calif., 1967.

4. *Temperature Stabilization of Klystron Oscillators by "Boiler Coolers,"* Varian Application Bulletin, AEB-20, Palo Alto, Calif., 1966.

5. Los, A., "Cooling High Voltage Power Supplies in Liquid Fluorochemicals," *Electronic Packaging and Production Magazine*, Aug., 1963, pp. 17–19.

Reference 1 provides the basic cooling theory from which the graphs and formulas of this chapter are derived.

Reference 2 describes the design of liquid filled electronics packages and also provides basic design equations for evaporation cooling.

Reference 3 describes the cooling of high power components by evaporation cooling.

Reference 4 discusses the use of evaporation cooling to provide constant temperature baths for electronic equipment, and includes a discussion of pressure equalization effects.

Reference 5 describes some practical liquid filled electronics packages.

7

Heat Pipes

Heat pipes are used in electronic equipment to "conduct" heat from the electronic component where the heat is generated over large distances to the cooling fins. Heat pipes are much more effective as heat conductors than the best metals. By using heat pipes to eliminate the problem of conducting heat to the fins, the fins can be made large enough to transfer the heat to the surroundings by the simplest means, such as radiation and natural convection or forced air cooling with low pressure fans. The use of heat pipes therefore greatly simplifies the overall cooling problem.

A schematic drawing of a heat pipe is shown in Figure 7.1. It consists of a hollow pipe which has been evacuated, filled with a coolant liquid, and sealed. The incident heat evaporates the liquid at one end of the heat pipe and the vapor transports the heat to the cooler end of the pipe. At the cooler end of the pipe the liquid condenses and transfers the heat out of the pipe. So far, the process is the same as evaporation cooling, which was discussed in Chapter 6. However, the heat pipe has an additional feature that permits it to be operated in any orientation, in contrast to evaporation cooling which can only work with the evaporator at the lowest point of the system. The inner surface of the heat pipe contains a capillary structure or "wick" which returns the condensed liquid to the hot evaporator end of the pipe by capillary action. The heat pipe can therefore operate in any orientation, even against gravity with its evaporator end upward.

A typical heat pipe for cooling electronic equipment would be a tube, 3/8 in. in diameter and 18 in. long. Such a heat pipe can conduct as much as 350 W from one end to the other with only a few degrees of temperature difference. By comparison, the temperature difference that would be required to conduct the same 350 W through 18 in. of a solid copper bar with the same 3/8 in. diameter would be 5700°C! (Of course, the copper bar would melt long before this temperature was reached.) Therefore, within its capa-

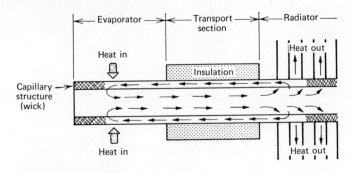

FIGURE 7.1
Schematic drawing of a heat pipe

bilities, a heat pipe can conduct several orders of magnitude more heat than a solid copper rod with the same cross-sectional area.

It is instructive to compare heat pipes to other methods of conducting heat from the electronic component to the cooling fins. This is done in Table 7.1, where the power handling capability, the possible orientations, and

TABLE 7.1
Comparison of Cooling Methods

Method	Power Handling Capability	Orientation	Necessary Auxiliary Equipment
Conduction through metals	Low	Any	None
Forced liquid cooling	High	Any	Pump, reservoir, interlocks, cooling lines
Evaporation cooling	Highest	One only	None
Heat pipe	High	Any	None

the necessary auxiliary equipment are compared for conduction through metals, forced liquid cooling, evaporation cooling, and heat pipes.

Table 7.1 clearly shows the advantages that the heat pipe offers over the other methods. The heat pipe has much greater heat conduction capability over long lengths than solid metal bars of the same cross section. The heat pipe has almost as high power handling capability as evaporation cooling and offers the important advantage that it will work in any orientation. Forced liquid cooling can provide as good power handling capability as heat pipes and can operate in any orientation but requires a considerable amount of auxiliary equipment, whereas the heat pipe requires none.

A typical design of the cooling of electronic equipment with heat pipes is described in Section 7.1. The design and fabrication of heat pipes are discussed in Section 7.2. The factors affecting heat pipe performance are described in Section 7.3, and an example of cooling with heat pipes is presented in Section 7.4 to illustrate the design information of this chapter.

7.1   A TYPICAL COOLING DESIGN USING HEAT PIPES

Figure 7.2 shows a typical electronic cooling design using heat pipes. The electronic components to be cooled are mounted on the heat sink in the center of the assembly. The heat generated by the components is conducted by the heat pipes to the cooling fins. When the fins are cooled with 80 CFM of air from a low pressure tubeaxial fan (such as the fan shown in Figure 4.13 of Chapter 4), 360 W can be dissipated by the assembly. Each of the four

FIGURE 7.2
Heat pipe—cooling fin assembly for electronic equipment (Photo courtesy of Noren Products)

heat pipes conducts 90 W from the mounting area to the fins with a temperature difference of only a few degrees. In contrast, if copper bars had been used instead of the heat pipes, the temperature difference between the mounting area and the fins would have been 245°C.

The use of heat pipes permits the fins on the assembly shown in Figure 7.2 to be made large enough so that they can be effectively cooled with a low pressure fan. In a conventional forced air cooled heat sink that does not use heat pipes, the cooling fins must be located close to the electronic component so that heat can be adequately conducted from the component to the fins. As a result, the size of the cooling fins is limited.

The advantage of using heat pipes is clearly illustrated by comparing the heat pipe-fin assembly shown in Figure 7.2 with the conventional forced air cooled heat sink described in Section 4.1 of Chapter 4. The conventional heat sink, which was shown in Figure 4.1, can dissipate almost as much power as the assembly with heat pipes, but only when a 400 cycle high pressure vaneaxial blower is used.

The low pressure tubeaxial fan that is used with the heat pipe-fin assembly can force only 18 CFM of air through the conventional forced air cooled heat sink and at this low air flow only 160 W can be dissipated. The use of heat pipes therefore doubles the cooling capability that can be obtained with a particular fan, or alternately, makes possible a given cooling capability with a fan which requires less power, is quieter, and has a much greater operating life.

7.2 HEAT PIPE DESIGN AND FABRICATION

The key to heat pipe operation is the capillary structure or wick which "pumps" the condensed liquid from the condenser back to the evaporator region. The capillary forces in the wick must be sufficient to overcome the viscous pressure drop in the liquid, the pressure of the evaporating vapor, and the force of gravity. The wick must achieve a balance between having large unobstructed flow paths to eliminate the viscous losses in the liquid and small pores to achieve large capillary forces. For these reasons, different wick geometries are used for different applications. Two of the commonly used wick geometries for heat pipes are shown in Figure 7.3.

A variety of fluids can be used in heat pipes. Typical fluids, and the operating temperature range of the heat pipes using them, are shown in Table 7.2.

The liquid metals are used in high temperature heat pipes for cooling nuclear reactor cores. For the temperature ranges normally used in electronic equipment, water and fluorochemical liquids, such as FC-43, are the most suitable fluids.

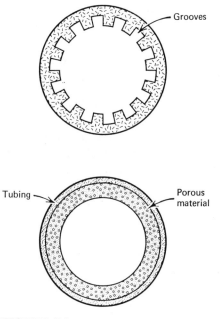

FIGURE 7.3
Wick designs for heat pipes

A heat pipe can operate satisfactorily over a wide range of temperature. For example, a water filled heat pipe can be used at any temperature from 10°C to 230°C. The temperature difference from the evaporator to the condenser end of the heat pipe is only a few degrees. The temperature of operation is determined by the temperature of the condenser end, which is in turn determined by the temperature of the cooling fins required to transfer the

TABLE 7.2
Heat Pipe Fluids

Fluid	Useful Temperature Range
Lithium	1000°C to 1700°C
Sodium	600°C to 1200°C
Cesium	400°C to 1000°C
FC-43	120°C to 220°C
Water	10°C to 230°C
Sulphur dioxide	−10°C to 43°C
Ammonia	−39°C to 22°C

heat to the surroundings. Operation over such a wide range of temperature is possible because the heat pipe is sealed. The internal pressure varies with the temperature of the condenser region. The boiling temperature of the fluid therefore varies to be always just a few degrees above the temperature of the condenser end of the heat pipe.

Heat pipes are as complicated to design and fabricate as the electronic components which they cool. The heat pipe must be designed to withstand the high internal pressures which result from operation over wide ranges of temperature. The wick must be designed to maximize capillary pumping and minimize viscous losses for the particular fluid used. The heat pipe must be carefully cleaned and evacuated to remove all impurities before filling with the working fluid.

The following companies specialize in heat pipe design and fabrication:

Energy Conversion Systems Research
P.O. Box 4208
Albuquerque, New Mexico 87106

Heat Pipe Corporation of America
141 Park Place
Watchung, New Jersey 07060

Isotopes, Inc. — A Teledyne Company
110 West Timonium Road
Timonium, Maryland 21093

Noren Products
3511 Haven Avenue
Redwood City, California 94062

Standard or specially designed heat pipes can be purchased from these heat pipe manufacturers, and then mounted in the electronic equipment. As always, care must be taken to minimize the temperature rise across the mounting interfaces.

The heat pipes shown in Figures 7.1 through 7.3 are all circular in cross section, and this is the most common configuration. However, heat pipes can be fabricated in a variety of shapes and cross sections to best meet specific cooling applications.

7.3   HEAT PIPE PERFORMANCE LIMITS

The maximum power that can be conducted by a heat pipe depends on:

• Evaporator area.
• Condenser area.
• Wick pumping capability.

An idealized drawing of a heat pipe is shown in Figure 7.4 to serve as an aid in estimating heat pipe performance limits. The outer diameter of this

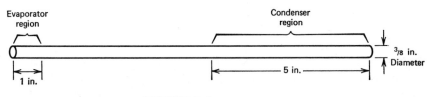

FIGURE 7.4
Typical heat pipe geometry

heat pipe is 3/8 in., and the inner diameter is .3 in. The evaporator area, where heat is applied to the heat pipe is 1 in. long, and the condenser area where heat is absorbed from the heat pipe is 5 in. long. The working fluid is water at 100°C. The internal evaporator area is .3 in. × π × 1 in., or approximately 1 in². As discussed in Chapter 6, the maximum heat flux that can be achieved by the evaporation of water at 100°C is 870 W/in². Therefore, the evaporator section of the heat pipe limits the power handling capacity to 870 W. The condenser area of this heat pipe is .3 in. × π × 5 in. or 5 in². As shown in Chapter 6, if the temperature difference between the vapor and the condenser wall is 10°C, the heat that can be transferred by the condensation process is 70 W/in². Thus the condenser section of the heat pipe limits the maximum power that can be handled to 70 W/in² × 5 in² = 350 W. Of course, the area of the condenser section could be increased, either by increasing the length or by increasing the diameter of the heat pipe, so that the condenser no longer limits the power capability.

The wick is usually the limiting factor in heat pipe power handling capability. The wick must have a large enough cross section to pump the total condensed liquid from the condenser back to the evaporator against the viscous forces of the liquid in the wick pores, against the vapor pressure of the evaporating liquid, and against the forces of gravity.

Table 7.3 shows the power capacity of typical commercially available heat pipes.

TABLE 7.3
Typical Heat Pipe Performance

Outside Diameter (inches)	Length (inches)	Power Capacity (watts)
1/4	6	300
3/8	6	500
1/2	6	700
1/4	18	150
3/8	18	350
1/2	18	550

Power capacity is for horizontal operation where the effects of gravity can be ignored. The above heat pipes operate over a temperature range from 20 to 200°C using water as the working fluid.

Although a heat pipe can operate in any orientation, its performance may be degraded when it must work against gravity, that is, when the evaporator region is higher than the condenser region. The effect of heat pipe orientation on the maximum power that can be conducted is shown qualitatively in Figure 7.5. In this figure the maximum power that can be conducted by the

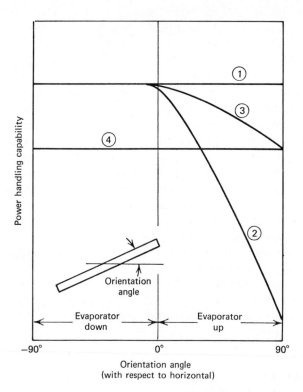

FIGURE 7.5
Heat pipe power capability as a function of orientation

heat pipe is shown as a function of the angle of elevation. At 0° the heat pipe is horizontal, at −90° the heat pipe is vertical with the evaporator end down. At 90° the heat pipe is vertical with the evaporator end up. Curve 1 shows the case of a heat pipe with a wick design which can pump sufficient fluid so that the heat pipe works equally well in any orientation. In this case the heat pipe performance is limited by the area of the condenser. Curve 2 shows a more typical case of an 8 in. long heat pipe where the power capacity is limited by the pumping capability of the wick. The power capacity is approximately constant as long as the evaporator end is lower than the condenser end. Even when this heat pipe is horizontal, its capacity is only slightly degraded. However, as the orientation is changed so that the evaporator end is up, the power capacity of the heat pipe becomes seriously degraded, decreasing by several times in the worst vertical orientation. Curve 3 shows a heat pipe with the same wick design but only half as long (4 in.). The length to which the wick must pump the liquid against gravity is shorter in this heat pipe, so its performance is not degraded so much in the worst vertical orientation. Curve 4 is the same shorter (4 in.) heat pipe, but with less of its length used for the condenser section. In this case the condenser section length and not the wick limits power capacity.

7.4 EXAMPLE

The use of the design information of this chapter for the cooling of electronic equipment with heat pipes is illustrated by the following example.

Figure 7.6 shows sketches of an electronic equipment chassis. The major heat generating component in the equipment is an SCR which dissipates 60 W. A conventional design for cooling this equipment is shown in Figure 7.6A. The SCR is mounted on a special heat sink so that the 60 W of heat can be transferred by radiation and natural convection, and the heat sink is then mounted on the top of the chassis.

A suitable heat sink for this application was described in Section 3.1. of Chapter 3 and was shown in Figure 3.1. When the heat sink is at 100°C, 60 W can be transferred to the surroundings which are at 20°C. This heat sink has a volume of 24 in^3, and as shown in the sketch of Figure 7.6A significantly increases the total volume of the equipment.

This conventional heat sink has a total surface area of 107 in^2. Because of the mutal shielding effects of the fins, less than half of this area is effective in transferring power by radiation.

The sides of the aluminum chassis have an even greater area than the finned heat sink. The area of three sides is 3 × 12 in. × 3 in. = 108 in^2. This area is not shielded and would therefore be an extremely effective radiating surface for transferring the 60 W of power dissipated by the SCR.

The only problem in using the sides of the chassis to transfer the heat generated by the SCR by radiation and natural convection is that the heat cannot be adequately conducted from the SCR through the 1/16 in. thick aluminum chassis.

A simple calculation using the formulas of Chapter 2 will show the problem of conducting the 60 W of heat through the sides of the chassis. Figure 7.6B

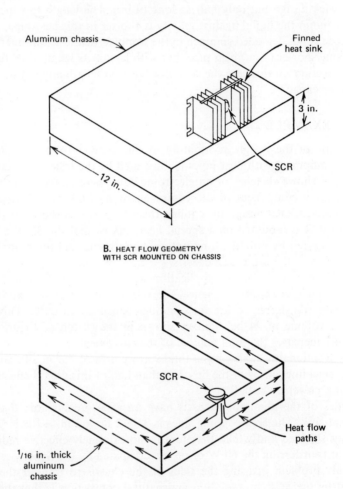

A. CONVENTIONAL DESIGN

Aluminum chassis

Finned
heat sink

3 in.

12 in.

SCR

B. HEAT FLOW GEOMETRY
WITH SCR MOUNTED ON CHASSIS

SCR

Heat flow
paths

1/16 in. thick
aluminum
chassis

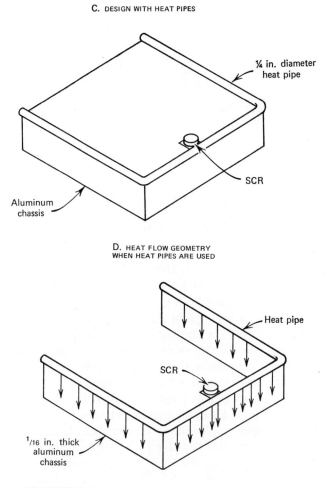

C. DESIGN WITH HEAT PIPES

¼ in. diameter
heat pipe

SCR

Aluminum
chassis

D. HEAT FLOW GEOMETRY
WHEN HEAT PIPES ARE USED

Heat pipe

SCR

$^1/16$ in. thick
aluminum
chassis

FIGURE 7.6
Use of heat pipes on an electronic equipment chassis

shows the geometry that will be used to simplify the calculation; conduction across the top of chassis is neglected.

Using Equation 2.3 from Chapter 2

$$\Delta T = \frac{1}{2} \frac{Q\lambda}{k\alpha}$$

$$= \frac{1}{2} \times \frac{30 \text{ W} \times 18 \text{ in.}}{5.5 \text{ W/in.}°\text{C} \times 1/16 \text{ in.} \times 3 \text{ in.}}$$

$$= 260°\text{C}$$

where:

Q (the power that must be conducted in one direction) is half of the total power or 30 W.

λ (the length through which the power must be conducted) is 18 in.

k (the thermal conductivity of aluminum) is 5.5 W/in.°C.

α (the cross-sectional area through which heat is conducted) is 1/16 in. $\times$ 3 in. $= .19$ in^2.

The factor of one-half which multiplies the equation is used to approximately account for the fact that the heat will be continuously transferred from the surface as it is conducted along the chassis sides. The factor of one-half is based on the assumption that the heat is transferred uniformly from the surface; actually the amount of heat transferred depends on the surface temperature which varies along the surface. For an accurate calculation, the fin efficiency formulas of Chapter 2 should be used, but use of more accurate formulas will not significantly change the result; the heat generated in the SCR cannot be adequately conducted along the aluminum chassis to utilize the surface of the sides of the chassis as a heat radiator.

The use of heat pipes readily solves the problem. Figure 7.6C shows a 1/4 in. diameter heat pipe mounted along the upper edge of the chassis. The heat pipe effectively distributes the 60 W of heat along the sides of chassis with a temperature drop of only a few degrees. The heat pipe adds an insignificant amount to the volume or weight of the chassis. As shown in Table 7.3, a 1/4 in. diameter heat pipe is more than adequate for conducting the 30 W of heat in each direction.

The heat pipe serves to distribute the heat along the entire upper edge of the chassis. Now the heat must only be conducted down the 3 in. height of the chassis sides as shown in Figure 7.6D. Using Equation 2.3, the approximate temperature difference required for conducting the heat down the sides of the chassis is

$$\Delta T = \frac{1}{2} \frac{Q\lambda}{k\alpha}$$

$$= \frac{1}{2} \times \frac{60 \text{ W} \times 3 \text{ in.}}{5.5 \text{ W/in.}°\text{C} \times 36 \text{ in.} \times 1/16 \text{ in.}}$$

$$= 7°\text{C}$$

where:

Q (the total power) is 60 W.
λ (the length through which heat is conducted) is 3 in.
k (the thermal conductivity of aluminum) is 5.5 W/in.°C.
α (the cross-sectional area through which heat is conducted) is 36 in. $\times$ 1/16 in. = 2.25 in^2.

The appropriate temperature difference is only 7°C between the top and bottom of the sides of the chassis.

Heat pipes can be fabricated in a variety of shapes and cross sections. For example, in this application the heat pipe could be square in cross section, 1/4 in. $\times$ 1/4 in., for easy mounting to the chassis.

7.5 REFERENCES

Useful references on heat pipes are as follows.

1. Eastman, G. Y., " The Heat Pipe," *Scientific American*, May, 1968, pp. 38–46.

2. *Can You Use a Heat Pipe*, Isotopes, Inc., Baltimore, Md., 1969.

3. Frank, S., *A Flat Plate Heat Pipe Heat Sink for High Power Electronic Components*, Isotopes, Inc., Baltimore, Md., Dec., 1968.

4. Basiulis, A., and Starr, M. C., " Improved reliability of TWT's through the use of a new lightweight heat removal device," *IEEE Transactions on Electron Devices*, **ED-15**, Aug., 1968, pp. 613–614.

Reference 1 describes the theory and operation of heat pipes and shows some typical applications.

Reference 2 describes typical heat pipe performance.

Reference 3 describes a particular application of heat pipes to the cooling of electronic equipment.

Reference 4 describes the application of heat pipe principles to the cooling of a high power microwave tube.

8

Refrigerated Equipment

The purpose of cooling electronic equipment is to keep the temperature of the electronic components at some desired temperature. With the methods of cooling discussed in all the previous chapters, the electronic component has always been hotter than the temperature of the surroundings. Heat flows from hot to cold bodies, so the electronic component had to be the hottest element in order to transfer its heat to the surroundings.

If necessary for achieving the desired electronic performance, the electronic component can be cooled to a temperature *below* the temperature of the surroundings by the use of refrigeration. In refrigerated equipment, heat does not flow from the electronic component, but is "pumped" by the refrigeration system from the cold component to the hot surroundings.

Examples of electronic equipment where refrigeration is required are:

1. Infrared detectors.
2. Masers.
3. Parametric amplifiers.
4. Equipment which must work in surroundings which are at a higher temperature than the safe operating temperature of the electronic components.

Refrigeration is not a substitute for good conventional cooling design. If the electronic component can be hotter than the surroundings, then by careful design, the heat can always be transferred from the component by conduction, radiation and natural convection, forced air cooling, forced liquid cooling, or evaporation cooling, and refrigeration is not necessary.

Refrigeration systems for electronic equipment may be classified into the following types:

1. Refrigerated cooling air or cooling liquid.
2. Refrigerated heat sinks.

204

3. Liquid nitrogen baths.
4. Thermoelectric coolers.

These refrigeration systems are discussed in Sections 8.1 through 8.4, respectively.

Figure 8.1 shows a schematic drawing of a refrigerated electronic equipment. The equipment consists of three basic parts:

1. A cold surface where the electronic component, which is at a temperature below the temperature of the surroundings, is mounted.
2. A hot surface where the heat transferred from the cold surface and the power dissipated by the refrigeration system are transferred to the surroundings. The hot surface must be at a higher temperature than the surroundings.
3. A refrigeration system, which pumps the heat from the cold surface to the hot surface.

The basic schematic of Figure 8.1 applies to any type of refrigerated electronic equipment, regardless of the type of refrigeration system used.

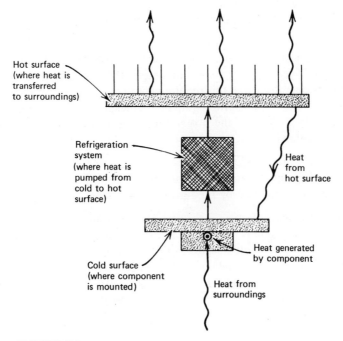

FIGURE 8.1
Schematic drawing of a refrigerated electronic component showing the heat flow paths

Figure 8.1 also shows the heat flow paths that exist in any refrigerated electronic equipment. The electronic component is the coldest element in the equipment, so heat flows back to the electronic component from the hot surface and the surroundings by radiation, natural convection and conduction. The total amount of heat that must be pumped from the cold to the hot surface therefore includes not only the heat generated by the electronic component but also this leakage heat from the surroundings and from the hot surface.

In order to reduce the size of the refrigeration system, every possible means should be used to thermally insulate the electronic component and the cold mounting surface from the hot surface and the surroundings. Useful means of insulation are:

1. Mounting the electronic component and the cold surface on poor thermal conductors, so heat cannot be conducted back from the hot surface and the surroundings.
2. Surrounding the cold surface with radiation shields so that the heat cannot be radiated back to the cold surface.
3. Evacuating the region around the component and the cold surface so that heat cannot be transferred back by natural convection.

Heat normally flows from a hot body to a cold body. In order to transfer heat in the opposite direction, work must be performed by the refrigeration system. A refrigerated electronic equipment therefore requires additional power for its operation, and this requirement of extra power for the refrigeration system reduces the overall equipment efficiency. This extra power ultimately appears as heat at the hot surface. The hot surface must therefore transfer the following heat to the surroundings:

1. The heat generated by the electronic component.
2. The leakage heat which flows back to the cooled surface.
3. The power dissipated by the refrigeration system.

In many refrigerated electronic equipments, the power required to operate the refrigeration system is more than an order of magnitude greater than the heat pumped from the cold surface to the hot surface. Therefore, the design of the hot surface to adequately transfer the heat to the surroundings must be carefully considered. The hot surface in any refrigerated equipment must always be hotter than the surroundings, and the heat may be transferred from it by radiation and natural convection, forced air, forced liquid, or evaporation cooling.

Specific types of refrigeration systems for electronic equipment will be described in Sections 8.1 through 8.4. The basic schematic diagram shown in Figure 8.1, which applies to any type of refrigerated electronic equipment, will be illustrated by these different types. In Section 8.5, an example of a refrigerated electronic equipment is presented to illustrate the design information of this chapter.

8.1   REFRIGERATION OF COOLING LIQUID OR AIR

A schematic diagram of a system for refrigerating the cooling liquid that is used in a forced liquid cooled electronics equipment is shown in Figure 8.2. A photograph of a typical system of this type is shown in Figure 8.3.

In a conventional forced liquid cooled electronics equipment, the components must be hotter than the cooling liquid so that they can transfer heat, and the cooling liquid must be hotter than the surroundings so that the heat

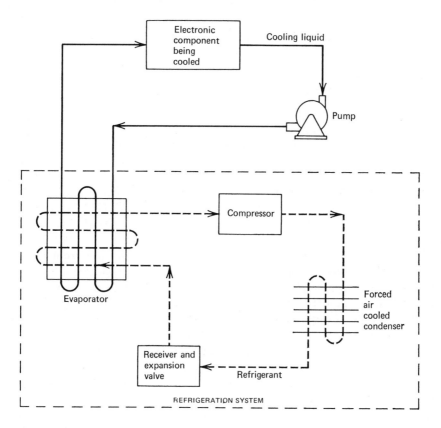

FIGURE 8.2
Block diagram of a refrigeration system for providing refrigerated cooling liquid to an electronic component

can be transferred from the liquid to the surroundings in an air-to-liquid heat exchanger before the liquid is recirculated through the equipment. Consequently, the electronics components are hotter than the surroundings.

In a refrigerated equipment, such as shown schematically in Figure 8.2, the components must still be hotter than the cooling liquid to transfer heat, but the liquid is refrigerated by the refrigeration system to a temperature below that of the surroundings, so the components themselves, although hotter than the cooling liquid, can be at a lower temperature than the surroundings. In the case of refrigerated equipment, the refrigeration system shown inside the dotted box in Figure 8.2 replaces the air-to-liquid heat exchanger of a conventional forced liquid cooled electronics equipment. The refrigeration system shown in Figure 8.2 is similar to that which would be found in a home refrigerator. This type of refrigeration system is described in detail in References 1 and 2, listed in Section 8.6. The refrigeration system consists of four basic parts:

1. Evaporator.
2. Compressor.
3. Condenser.
4. Receiver and expansion valve.

The cooling liquid and the refrigerant liquid both circulate through the evaporator in separate lines. The refrigerant liquid, which is usually a Freon, absorbs heat from the cooling liquid and evaporates. This heat absorption process is the same as the evaporation cooling described in detail in Chapter 6. The boiling point of the Freon refrigerant is less than 0°C. (The exact value depends on the refrigeration system pressure.) As heat is absorbed from the cooling liquid, its temperature is lowered so that it can enter the electronics equipment at a temperature well below the temperature of the surroundings.

The low pressure, low temperature refrigerant vapor which results from the evaporation process is sucked from the evaporator to the compressor where it is compressed to a high pressure, high temperature vapor. The temperature of the vapor is raised in the compressor above the temperature of the surroundings. The compressor does work on the vapor and this work appears as heat energy. The high pressure refrigerant vapor then flows to the condenser where the vapor condenses to a liquid at the same high temperature and gives off its heat to the surroundings. The heat transferred to the surroundings in the condenser includes both the heat absorbed from the coolant liquid and the work supplied by the compressor. The high pressure refrigerant liquid then flows from the condenser to the receiver and through an expansion valve where the pressure of the liquid is lowered, and back to the evaporator where the refrigeration cycle begins again.

FIGURE 8.3
Typical refrigeration system for providing refriger-
ated cooling liquid to an electronic equipment
(Photo courtesy of Electro Impulse Laboratory,
Inc.)

A typical refrigeration system for supplying refrigerated cooling liquid to
electronics equipment is shown in Figure 8.3. This system has all the elements
shown in the schematic diagram of Figure 8.2, and in addition has controls
and metering for the cooling liquid. The system shown in Figure 8.3 can
provide 1 gal/min of refrigerated water at 60 psi. When the surroundings are
at 25°C, the system can maintain the inlet cooling water at 5°C when the
power absorbed by the water is 1250 W. If more power must be absorbed, the
water cannot be maintained at as low a temperature. For example, if 2500 W
must be absorbed, the minimum water temperature is 25°C. In this case, the
coolant water is at the same temperature as the surroundings. However, if a

refrigerated system were not used, the water would have to be at a higher temperature than the surroundings.

The refrigeration and water circulation system shown in Figure 8.3 is 17 1/2 in. × 24 in. × 28 in. high. It fits into a standard relay rack and weighs 260 lb.

It is instructive to compare the performance specifications of the refrigerated liquid cooling system shown in Figure 8.3 with the conventional nonrefrigerated forced liquid cooling system which was described in Chapter 5 and was shown in Figure 5.17. Both systems are made by the same manufacturer and both systems provide 1 gal/min of cooling water at 60 psi. The comparative performance of these two systems is shown in Table 8.1.

TABLE 8.1

Comparative Performance of Refrigerated and Conventional Forced Liquid Cooling Systems

Specification	Conventional Forced Liquid Cooling System	Refrigerated Forced Liquid Cooling System
Water temperature when power absorbed is 1250 W	31°C	5°C
Water temperature when power absorbed is 2500 W	36°C	25°C
Water temperature when power absorbed is 5000 W	50°C	—
Size	17 1/2 in. × 24 in. × 14 in.	17 1/2 in. × 24 in. × 28 in.
Weight	75 lb	260 lb

The water temperatures given in Table 8.1 are for the case when the temperature of the surroundings is 25°C. As the table clearly shows, the refrigerated system occupies two times the volume and weighs 3 1/2 times as much as the nonrefrigerated system. At low power levels of 1250 W, the refrigerated cooling system can reduce the coolant temperature well below the temperature that can be obtained with the nonrefrigerated system. However, when the power absorbed is 2500 W, the advantage of the refrigerated system is negligible. Also the smaller, lighter conventional cooling system can handle considerably more power than the refrigerated system.

The refrigerated coolant system such as shown in Figure 8.3 can also be provided with a heater and thermostat in the liquid cooling line so that the inlet temperature of the coolant can be maintained constant, independent of the temperature of the surroundings, by using either the refrigeration system or the heater.

The problem of humidity must be carefully considered when using refrigerated cooling liquid in electronic equipment. If the electronic components are cooled below the temperature of the surroundings, then moisture in the surrounding air can condense on the components. This condensation can cause shorting or arcing; therefore, if refrigerated cooling liquid is used, the component mounting must be designed to protect exposed voltages from condensation.

FIGURE 8.4
Refrigeration system for providing refrigerated cooling air to a rack of electronic equipment (Photo courtesy of Electro Impulse Laboratory, Inc.)

Figure 8.4 shows a refrigerated cooling air system. The system is similar to the refrigerated cooling liquid system which was shown schematically in Figure 8.2. In this case, the refrigeration unit mounted in the bottom of the rack is used to refrigerate the cooling air which is then circulated over the electronic components which are mounted in the rack.

8.2   REFRIGERATED HEAT SINKS

Even lower component temperatures than can be obtained by refrigerating the cooling liquid or cooling air can be obtained by using refrigerated heat sinks. A schematic drawing of a refrigerated heat sink is shown in Figure 8.5. In this case, the evaporator of the refrigeration system is the heat sink itself, and the electronic components are mounted directly on the outer

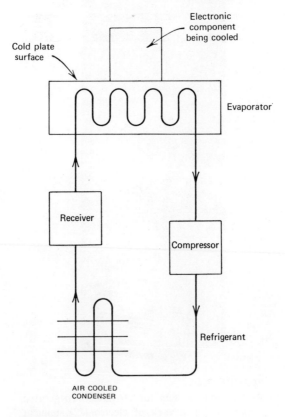

FIGURE 8.5
Block diagram of a refrigerated heat sink

surface of the evaporator. In the refrigerated liquid cooling system described in Section 8.1, the refrigerated liquid had to be hotter than the evaporator to transfer heat to it and the electronic components had to be hotter than the coolant liquid. With the refrigerated heat sink, the intermediate step of the coolant liquid can be eliminated, and the electronic component temperature can approach the temperature of the evaporator itself.

A typical Freon refrigeration system which can lower the heat sink temperature to $-25°C$ operates at internal pressures as high as 150 psi. Consequently, the evaporator heat sink and all refrigerant lines and fittings which connect the refrigeration system to the heat sink must be designed to withstand high pressure. In this respect, much more care must be taken in integrating the refrigeration equipment into the electronics equipment with a refrigerated heat sink than in the case of the refrigerated coolant liquid or air systems described in the previous section.

As with all refrigerated electronic equipment, the problem of thermal insulation of the refrigerated heat sink and the problem of condensation of liquid from the surrounding air must be solved.

By replacing the conventional Freon refrigeration system shown schematically in Figure 8.5 with a cryogenic refrigeration system, temperatures down to $-250°C$ can be obtained. The details of the cryogenic refrigeration equipment are beyond the scope of this book, but are described in detail in References 3 and 4.

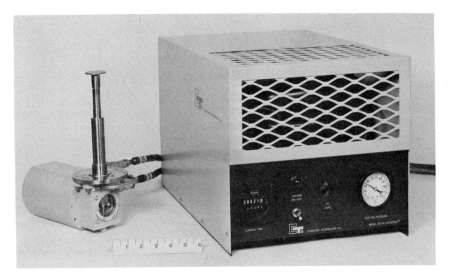

FIGURE 8.6
Typical cryogenic cooled heat sink (Photo courtesy of Cryogenic Technology, Inc.)

Figure 8.6 shows a typical cryogenic heat sink system. This system can cool a heat load of 1 W down to $-250°C$, or can cool a heat load of 10 W down to $-195°C$. The electronic component to be cooled is mounted on the small 1 in. diameter cylindrical plate on the top of the column. To provide adequate insulation at these low temperatures, the component and cold plate should be mounted in a vacuum. The vacuum jar is not shown in the figure, but seals to the lower cylindrical plate. The system shown requires electrical power only. No cold liquids or external gases are required. The complete system weighs 100 lb, and requires 1500 W of 110 V 60 cycle AC power. Cooldown time is less than 30 min from room temperature to $-250°C$.

8.3 LIQUID NITROGEN BATHS

One of the simplest methods of cooling electronic components below the temperature of the surroundings is to immerse them in a liquid nitrogen bath. This cooling technique, which is shown schematically in Figure 8.7, is just the same as the evaporation cooling described in Chapter 6, except that in this case, the boiling point of the liquid is below the temperature of the surroundings instead of above it. Any liquified gas can be used as the cooling liquid, but because of reasons of toxicity, flammability and availability, liquid nitrogen is most commonly used.

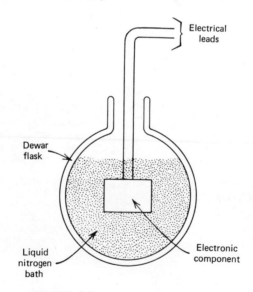

FIGURE 8.7
Schematic drawing of a liquid nitrogen bath for cooling electronic equipment

The boiling point of nitrogen is $-195°C$ at atmospheric pressure. Its heat of vaporization is 23 W hr/lb or approximately .08 that of water.

Liquid nitrogen cooling baths are usually used as expendable evaporation cooling systems. The liquid nitrogen absorbs heat and evaporates and the resulting vapor is expended to the surroundings. The liquid nitrogen must therefore be periodically replaced in the system. It is usually more economical to liquify nitrogen in large liquification systems, and transport it in liquid form to the electronics equipment, than to provide a closed cycle liquification system for an individual electronic equipment.

A typical example of a liquid nitrogen cooled electronics equipment is shown in Figure 10.10 of Chapter 10. This equipment is an infrared detector used for measuring the temperature distribution in electronic equipment. The infrared detector must operate at low temperature, and is cooled to the required low temperature by immersion in liquid nitrogen.

The schematic diagram of an electronic component cooled in a liquid nitrogen bath shown in Figure 8.7 shows the critical elements that must be considered in designing this type of cooling system. The liquid nitrogen bath must be insulated from the surroundings so that heat leakage from the warmer surroundings to the cooler bath is minimized. This insulation is most commonly obtained using a Dewar flask. A Dewar flask is a double walled container made of a thermal insulating material such as glass. The space between the walls is evacuated to minimize heat conduction and both surfaces are coated with a shiny metal film to minimize radiation losses. Special care must also be taken to thermally insulate the electrical leads to the component to minimize conductive heat losses through the leads. Even when these precautions of insulation are taken, the major fraction of the heat which is absorbed by the evaporating liquid nitrogen is leakage from the surroundings. In most cases the power dissipated by the electronic component itself is negligible compared to this leakage heat, so the minimizing of this leakage heat by the use of insulation techniques is extremely important.

8.4   THERMOELECTRIC COOLING

Thermoelectric cooling provides one of the simplest means of refrigerating electronic equipment. The thermoelectric cooling module operates on electronic principles so that the refrigeration system requires no compressors, evaporators, condensers, refrigerant liquids, etc. Thermoelectric cooling modules can pump as much as 100 W of heat from a few degrees below the temperature of the surroundings or can reduce the component temperature to $-100°C$ below the surroundings if only a few milliwatts of heat must be pumped.

A schematic drawing of a thermoelectric cooler is shown in Figure 8.8. The thermoelectric cooler consists of a type N and a type P semiconductor of bismuth telluride. A junction between these dissimilar semiconductors is formed at the surface to be cooled and a DC voltage is applied across the other junction at the hot surface where heat is transferred to the surroundings. The extra electrons in the N type material and the holes left in the P type

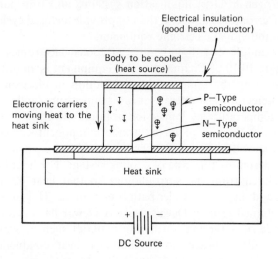

FIGURE 8.8
Schematic drawing of a thermoelectric cooling element

material are the carriers that move the heat energy from the cold to the hot junction. The heat is pumped by virtue of the Peltier effect. Heat absorbed at the cold junction is pumped to the hot junction at a rate proportional to the carrier current passing through the circuit. Good thermoelectric semiconductor materials such as bismuth telluride impede the conventional conduction of heat from the hot to the cold surface, yet provide an easy flow for the carriers which are transferring heat in the opposite direction.

Thermoelectric elements are usually connected in parallel to obtain the required power handling capacity. A drawing of a typical module assembly with the individual elements connected thermally in parallel and electrically in series is shown in Figure 8.9.

A typical thermoelectric cooling module is shown in Figure 8.10. This module is 1 5/32 in. × 1 5/32 in. × 7/32 in. high. It consists of many individual thermoelectric elements in parallel as shown schematically in Figure 8.9, and can pump 5 W of heat from the cold to the hot junction through a 50°C

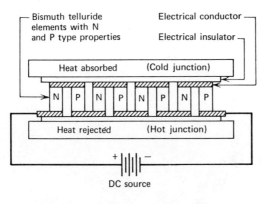

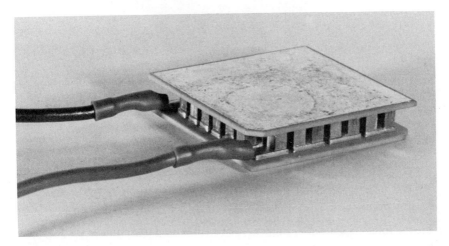

FIGURE 8.9
Schematic drawing of a typical thermoelectric module assembly

FIGURE 8.10
Typical thermoelectric cooling module (Photo courtesy of Borg-Warner Corp.)

temperature difference. If the hot junction is at 25°C, then the 5 W of heat can be pumped from a cold junction which is at −25°C. At other power levels the temperature difference between the hot and cold junction is different, and these tradeoffs are discussed in a subsequent paragraph.

Figure 8.11 shows a complete thermoelectric cooler with an air cooled heat sink. The air cooling system must be designed to transfer:

FIGURE 8.11
Complete thermoelectric cooler with an air cooled heat sink (Photo courtesy of Borg-Warner Corp.)

1. The power dissipated by the electronic component.
2. The leakage heat which flows from the hotter surroundings to the cooled component.
3. The power expended in pumping the heat from the cold to the hot junction.

Note the foam insulation in Figure 8.11 which is used to minimize the leakage heat.

The temperature distribution through a thermoelectric cooling unit is shown in Figure 8.12. The heat generated by the electronic component must first be conducted through an alumina ceramic insulator and then through a metal contact which makes the electrical connection between the N and P type semiconductor materials. In order that this heat can be conducted from the component to the cold junction, the component must be hotter than the junction, and this temperature difference through the ceramic insulator and through the metal contact is shown in Figure 8.12. The cold junction itself is the coldest point in the thermoelectric cooling system. Once the heat is

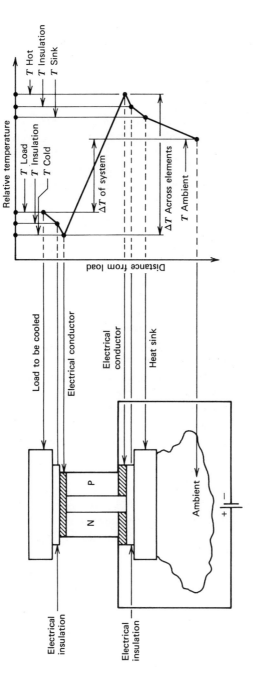

FIGURE 8.12
Temperature distribution through a thermoelectric cooling unit

pumped to the hot junction, it must be conducted through another metal contact, through another ceramic insulator, and then be transferred to the surroundings by radiation and natural convection or forced air cooling. Consequently, the temperature difference between the surroundings and the cooled electronic component is less than the total temperature difference through which heat can be pumped by the thermoelectric element. For this reason, it is important to have a good thermal conduction path from the component to the cold junction and from the hot junction to the hot surface where heat is transferred to the surroundings. It is also important to design the heat transfer from the hot surface so as to minimize the temperature difference between the surface and the surroundings to take full advantage of the cooling capability of the thermoelectric element.

The performance of the typical thermoelectric cooling module shown in Figure 8.10 is illustrated in Figure 8.13, where the temperature difference between the cold surface where the electronic component is mounted and the heat sink mounting surface is shown as a function of the power that is transferred. The data shown in Figure 8.13 takes into account the temperature rise required to conduct heat from the component mounting surface to the cold junction and from the hot junction to the heat sink mounting surface. The heat pumping capacity of any thermoelectric module depends on the amount of current that is driven through the module. This is illustrated for the particular thermoelectric cooling module in Figure 8.13 by the three lines at constant currents of 2 A, 4 A, and 8.5 A. With any given current, the greater the heat that is pumped from the cold to the hot surface, the less the temperature difference that can be maintained between the surfaces. For example, in Figure 8.13, considering the case where the thermoelectric cooling module is operated at its maximum current rating of 8.5 A, if the heat pumped from the cold to the hot surface is negligible, the temperature difference between the cold and hot surface can be as great as 70°C. If the heat pumped from the cold to the hot surface is 10 W, then the maximum temperature that can be maintained between the cold and hot surface is only 35°C, and if the power that is pumped between the cold and hot surface is near the maximum capability of 19 W, the temperature difference can be only a few degrees.

The voltage that must be applied to drive the current through the thermoelectric module is a nonlinear function of the current. At 8.5 A of current, the voltage is approximately 3.7 V, and therefore the total power that is required to operate the thermoelectric module is 32 W. A useful measure of the efficiency of a thermoelectric cooling module is the "coefficient of performance," which is the ratio of the heat pumped from the cold to the hot junction to the power that must be supplied to operate the module. Two values of coefficient of performance are plotted in Figure 8.13. For example, assuming that the current through the thermoelectric module is 8.5 A and

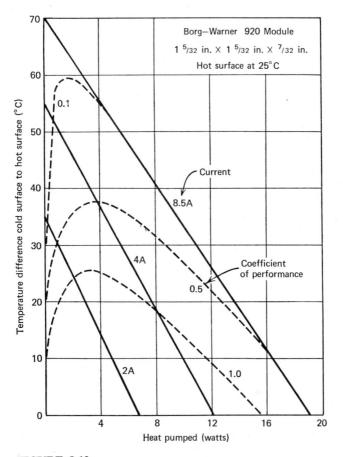

FIGURE 8.13
**Temperature difference from the cold surface to the hot surface
as a function of heat pumped for the thermoelectric cooling
module shown in Figure 8.10**

16 W of power is pumped from the cold to the hot surface, then the coefficient of performance is .5. In this case the hot surface must transfer both
the 16 W of heat which is pumped from the cold to the hot surface and the
32 W of power which is required to operate the thermoelectric element (which
appears as heat at the hot surface) for a total of 48 W. Note that the 16 W
pumped from the cold to the hot surface includes both the power that is
dissipated by the electronic component and the leakage heat which flows
from the surroundings to the cold surface.

The data shown in Figure 8.13 is for the case when the hot surface is at
25°C. If the hot surface is at other temperatures, the curves are different. The

insections of the curves with both the temperature difference and heat pumped axes increase at the rate of $1/2\%/°C$ temperature increase in hot surface temperature. This behavior is shown more specifically in Figure 8.14.

If greater power must be pumped from the cold to the hot surface than can be obtained with the module shown in Figure 8.10, a larger module with more elements in parallel can be used. Such a larger module was shown in Figure 8.11 with its air cooled heat sink. The performance of this larger, higher capacity module is shown in Figure 8.14. In this case, all the curves are for the maximum operating current of 6.5 A, but different curves are shown for various temperatures of the surroundings. At the highest surrounding temperature of 75°C, much more power can be pumped from the cold to the hot surface for a given temperature difference between the hot and cold surface. Note that with this larger unit, which requires 162 W of power for its operation (i.e., 6.5 A at 25 V), 40 W of heat can be transferred from the cold surface at 0°C to the hot surface at 25°C.

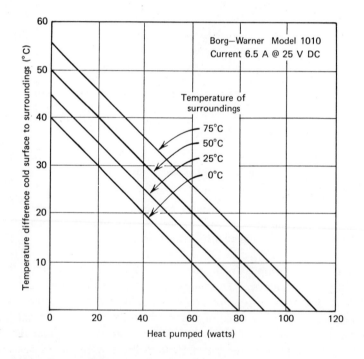

FIGURE 8.14
Temperature difference between the cold surface and the surroundings as a function of heat pumped for the thermoelectric cooler shown in Figure 8.11

FIGURE 8.15
Four stage cascaded thermoelectric cooler (Photo courtesy of Borg-Warner Corp.)

When the required temperature difference between the cold and the hot surface is greater than can be achieved with a single thermoelectric module, the modules can be cascaded in series. A typical four stage cascaded thermoelectric cooler is shown in Figure 8.15, and the performance of this particular unit is shown in Figure 8.16. Each successive higher temperature stage must pump not only the heat generated by the electronic component but also the power required for the operation of the lower temperature module. Consequently, the highest temperature stage must pump more than an order of magnitude more power from its cold to its hot surface than the lowest temperature stage. This requirement accounts for the pyramid type appearance of the cascaded module, and it also limits its power handling capability to less than a watt, as shown in Figure 8.16. The performance of this particular module also shows the importance of thermally insulating the module so that heat will not flow back from the hot to the cold surface. Note that when this particular module is operated in a vacuum with the hot surface at 25°C, the cold surface can be at −90°C for a total temperature difference

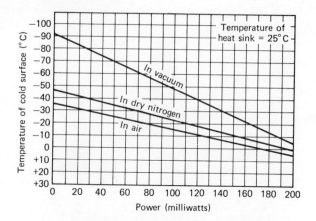

FIGURE 8.16
Temperature of the cold surface as a function of heat pumped for the four stage cascaded thermoelectric cooler shown in Figure 8.15

between the hot and cold surface of 115°C when low powers of a few milliwatts are pumped. If, however, this same module is mounted in air, then the temperature difference between the cold and hot surface is only 60°C. The obtainable temperature difference is reduced almost in half. The reason for this, of course, is that approximately 125 mW of power leaks from the hot to the cold surface by natural convection, and this leakage power must also be pumped from the cold to the hot surface. The module is shown in Figure 8.15 mounted on a vacuum flange, and the performance curve shows clearly the advantage of the vacuum insulation.

8.5 EXAMPLE

The use of the information of this chapter for designing refrigerated electronic equipment is illustrated by the following example.

A microwave receiver which uses a parametric amplifier must be cooled to below 0°C to obtain optimum performance. It can be cooled to this temperature by the thermoelectric cooling module shown in Figure 8.10.

The temperature sensitive elements of the parametric amplifier are mounted on the 1 5/8 in. × 1 5/8 in. cold surface of the thermoelectric module. Two watts of power are dissipated by the amplifier. In addition to this power, leakage heat is transferred from the surroundings to the cold surface by radiation and convection and by conduction through the electrical leads to the amplifier. The greater the temperature difference between the hot surface and the cold surface, the greater the amount of leakage power. The leakage power is approximately .02 Wper°C temperature difference between the hot and cold surface. This leakage power can be calculated from the formulas of Chapter 2 for heat conduction and of Chapter 3 for radiation and natural convection.

The first step in designing this refrigerated electronic equipment is to determine the minimum temperature that can be obtained with the thermoelectric cooling module. The performance of this module was shown in Figure 8.13 when the hot surface was at 25°C. If the surroundings are at 25°C, the hot surface must be at a higher temperature in order to effectively transfer heat to the surroundings. In this example it will be assumed that the hot surface is at 65°C and that the temperature difference of 40°C between the hot surface and the surroundings is necessary to transfer heat from the hot surface to the surroundings. The performance of the module is replotted in Figure 8.17 when it is operated at its maximum current rating of 8.5 A and the applied voltage is 3.7 V and the hot surface is at 65°C. The curve shown in Figure 8.17 was constructed from the standard performance curve of Figure 8.13 by using the fact that the intersections on the temperature difference and heat pumped axes change by approximately $1/2\%/°C$ increase in hot surface temperature.

The dotted line on Figure 8.17 shows the leakage power. The slope of this line is .02 W/°C. The solid line shows the combination of this leakage power and the power dissipated by the amplifier of 2 W, and this total power must be pumped from the cold to the hot surface. This load line intersects the characteristic curve for the thermoelectric module at a temperature difference of 70°C and a power pumping capability of 3.5 W. Therefore, the thermoelectric cooling element can maintain a temperature difference of 70°C between the hot and the cold surface when pumping 3.5 W of power. Two watts of this power come from the parametric amplifier itself and the other 1.5 W is leakage power from the hot surface and the surroundings back to the cold surface. Since the hot surface is at 65°C, the cold surface can be maintained at $-5°C$ by the thermoelectric cooling module.

The final step is to determine the heat that must be transferred from the hot surface to the surroundings. This heat includes the power dissipated by the parametric amplifier, the leakage power, and the power supplied to the thermoelectric module. These powers are as follows.

Power dissipated by electronic component	2.0 W
Leakage heat from surroundings	1.5 W
Power required to operate thermoelectric cooler	31.5 W
Total Power	35 W

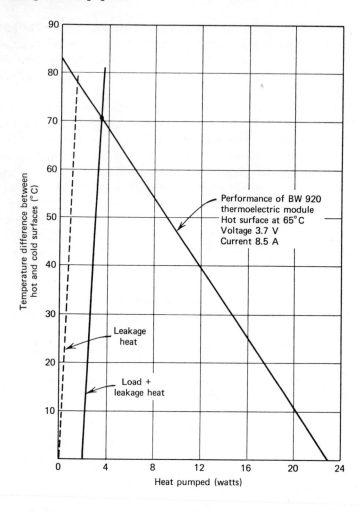

FIGURE 8.17
Temperature difference between the hot and cold surfaces of a thermoelectric cooling module as a function of heat pumped

The hot surface must be designed to transfer this 35 W of power to the surroundings at a 40°C temperature difference between the hot surface and the surroundings. The hot surface may be cooled by radiation and natural convection, forced air, or any of the other means of transferring heat from a hotter surface to the surroundings.

8.6   REFERENCES

Useful references for refrigerated electronic equipment are as follows.

1. *Introduction to Cooling Units*, Technical Bulletin 696, Electro Impulse, Inc., Red Bank, N.J., 1969.

2. Manley, H. P., *Drakes Refrigeration Manual*, Frederick J. Drake, Chicago, 1962.

3. Croft, A. J., *Cryogenic Laboratory Equipment*, Plenum Press, New York, 1970.

4. Bell, J. H., Jr., *Cryogenic Engineering*, Prentice Hall, Inc., Englewood Cliffs, N.J., 1963.

5. Boesen, G. F., *et al.*, *The Where and the Why of Thermoelectric Cooling*, Borg-Warner Corporation, Des Plaines, Ill., 1967.

6. Krauss, A. D., *Cooling Electronic Equipment*, Prentice Hall, Inc., Englewood Cliffs, N.J., 1965, Chapter 15, pp. 294–326.

Reference 1 describes refrigeration systems for providing refrigerated cooling liquid or refrigerated cooling air for use with electronic equipment.

Reference 2 is a basic refrigeration manual, and describes refrigeration systems in general.

References 3 and 4 describe cryogenic refrigeration equipment and treat in detail the problems of insulation, storage, and handling of liquid nitrogen.

Reference 5 discusses commercially available thermoelectric cooling modules for electronic equipment, including their performance and application.

Reference 6 discusses the theory of thermoelectric coolers and the various physical factors limiting their performance.

9

Transient Effects

The previous chapters have all discussed steady state conditions. Under steady state conditions, heat is transferred to the surroundings as fast as it is generated by the electronic component, and the temperatures of all elements—the component itself, the heat sink, the cooling fins, and the cooling air or liquid—remain constant with time.

However, when electronic equipment is first turned on, or when the power input or the coolant flow rates are changed, the temperature of the electronic component and all other elements in the cooling system vary with time, until the equipment reaches its steady state temperature. Because many electronic components change their electrical characteristics when their temperature changes, the transient thermal characteristics of the equipment must be considered.

Design information on transient thermal conditions in electronic equipment is presented in this chapter. The warmup characteristics of a typical heat sink are described in Section 9.1. Design equations for calculating transient thermal conditions are presented in Section 9.2. Examples of transient cooling of electronic equipment are calculated in Section 9.3 to illustrate the equations and tables of this chapter.

9.1 WARMUP CHARACTERISTICS OF A TYPICAL HEAT SINK

A typical heat sink for electronic equipment is shown in Figure 9.1. This same heat sink, which is cooled by radiation and natural convection, was discussed in Section 3.1 of Chapter 3, and its steady state thermal performance was calculated in Example 1 of Section 3.4.

228

FIGURE 9.1
Heat sink cooled by radiation and natural convection
(Photo courtesy of Wakefield Engineering, Inc.)

When the electronic component that is mounted in the heat sink dissipates 60 W and the surroundings are at 20°C, the steady state temperature of the heat sink is 100°C.

The warmup characteristics of this heat sink are shown in Figure 9.2. Figure 9.2A shows the power dissipated by the electronic component, the power transferred to the surroundings, and the net power which heats up the heat sink, all as a function of time.

Figure 9.2B shows, on the same time scale, the temperature of the heat sink.

Before the equipment is turned on, the heat sink and the electronic component are at the temperature of the surroundings of 20°C. When the equipment is turned on, the electronic component immediately begins to generate 60 W of heat, and as shown by Figure 9.2A, this generated power remains constant with time. Just after turnon, all of the 60 W are effective in raising the temperature of the heat sink.

As the heat sink rises in temperature, it begins to transfer heat to the surroundings, and then, as shown in Figure 9.2A, only part of the total heat generated by the electronic component is effective in raising the temperature of the heat sink. For example, at 5 min after turnon, the heat sink has risen to 63°C, which is 43°C above the temperature of the surroundings. At this temperature, 33 W are being transferred to the surroundings, and only 27 W

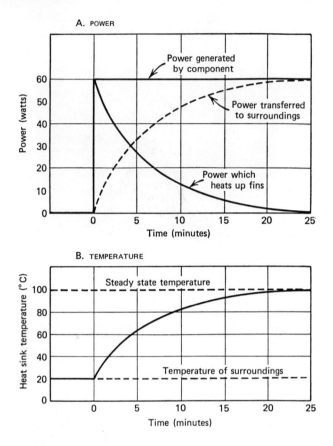

FIGURE 9.2
Warmup characteristics of the heat sink shown in Figure 9.1

are effective in raising the temperature of the heat sink. As time goes on, the heat sink temperature increases, more of the total heat generated by the component is transferred to the surroundings, less is available to raise the temperature of the heat sink, and, as shown by Figure 9.2B, the rate of change of temperature continually decreases. After 15 min, the heat sink has reached 90% of its steady state temperature, and 55 W of the total of 60 W are being transferred to the surroundings with only 5 W being effective in raising the temperature of the heat sink.

After a long enough time has elapsed, the heat sink will finally reach the steady state condition where heat is transferred to the surroundings as fast

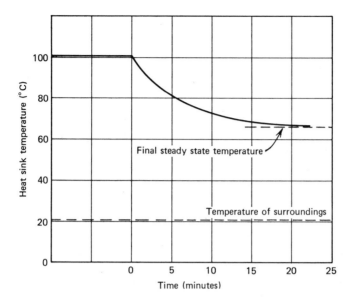

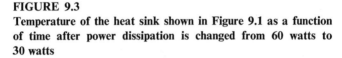

FIGURE 9.3
Temperature of the heat sink shown in Figure 9.1 as a function
of time after power dissipation is changed from 60 watts to
30 watts

as it is generated by the component. At this time, the heat sink temperature
is 100°C.

Another interesting transient thermal characteristic of this heat sink is
illustrated in Figure 9.3. The heat sink has been operated for a long time
with the electronic component dissipating 60 W, and the heat sink has
reached its steady state temperature of 100°C. Suddenly, the power generated
by the electronic component is reduced to 30 W. At this time, more heat is
being transferred to the surroundings than is being generated by the component, and the heat sink temperature begins to decrease. This cooling
process is just the reverse of the warmup process described previously.
Finally, after more than 20 min, the heat sink again reaches a steady state
condition where the 30 W is transferred to the surroundings just as fast as
it is generated, and the heat sink reaches a steady state temperature of 66°C.

The transient characteristics of this particular heat sink are calculated in
Example 1 of Section 9.3 to illustrate the use of the design information of
this chapter.

9.2   CALCULATION OF TRANSIENT EFFECTS

The transient temperature of any heat sink can be calculated from the following formula:

$$\Delta T(\gamma) = \Delta T(\infty) - [\Delta T(\infty) - \Delta T(0)]e^{-\gamma/(\Delta T/Q)cM} \qquad (9.1)$$

where:

$\Delta T(\gamma)$ is the transient temperature difference between the heat sink and the surroundings at any time γ (°C).

$\Delta T(0)$ is the temperature difference between the heat sink and the surroundings when the transient starts, i.e., when $\gamma = 0$ (°C).

$\Delta T(\infty)$ is the final steady state temperature difference between the heat sink and the surroundings, i.e., when $\gamma = \infty$ (°C).

$(\Delta T/Q)$ is the steady state temperature difference between the heat sink and the surroundings required to transfer a given amount of power (°C/W).

γ is the time after the transient starts (minutes).

c is the specific heat of the heat sink (watt-minute)/(pound-°C).

M is the weight of the heat sink (pounds).

e is 2.718, the base of natural logarithms.

Equation 9.1 can be rearranged into the following form.

$$\frac{\Delta T(\gamma) - \Delta T(\infty)}{\Delta T(0) - \Delta T(\infty)} = e^{-\gamma/[(\Delta T/Q)cM]}. \qquad (9.2)$$

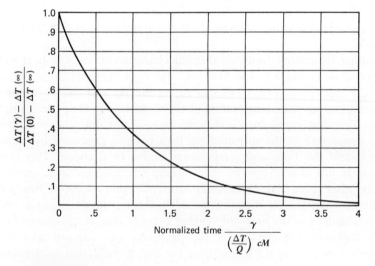

FIGURE 9.4
Graphical presentation of Equation 9.2

Equation 9.2 shows the difference between the heat sink temperature at any time and the final steady state temperature, relative to the difference between the initial and final temperatures.

Equation 9.2 is shown graphically in Figure 9.4 as a function of normalized time $\gamma/[(\Delta T/Q)cM]$.

A particularly interesting time is when the temperature of the heat sink is within 5% of its steady state value. From Figure 9.4, or Equation 9.1, this occurs when

$$(\gamma)_{5\%} = 3\left(\frac{\Delta T}{Q}\right)cM. \tag{9.3}$$

The time required for the heat sink temperature to come within 5% of its steady state value, as given by Equation 9.3, or any particular other percentage, is proportional to:

1. The heat transfer capability of the heat sink $(\Delta T/Q)$.
2. The specific heat of the material of which the heat sink is made (c).
3. The weight of the heat sink (M).

The greater the specific heat of the material, the greater the weight of the heat sink, or the poorer the heat transfer capability of the heat sink, the longer is the time required for the heat sink to reach a given fraction of its steady state temperature. The heat transfer capability of the heat sink and the steady state temperature difference between the heat sink and the surroundings can be calculated from the design information of Chapter 3, 4, 5, or 6, depending on whether the heat sink is cooled by radiation and natural convection, forced air cooling, forced liquid cooling, or liquid evaporation.

The value of specific heat and the weight of the heat sink can be determined from Table 9.1, which shows the specific heat and the density of commonly used heat sink materials.

A typical heat sink for cooling electronic components will usually consist of several parts, for example, the component itself, the insulator, and the heat sink. Each of these individual parts has a different weight and a different specific heat. For use with Equations 9.1 through 9.3, it is usually satisfactory to include only the heat sink itself, since it contributes the major part of the weight.

The best way to determine the weight of the heat sink is to weigh the actual part. Because this is not possible when the heat sink is being designed, the weight must be calculated from the volume of the heat sink and the density of the material from which it is made. The density of common heat sink materials is therefore given in Table 9.1.

Equations 9.1 through 9.3 neglect transient effects involving the conduction of heat through the heat sink to the cooling fins. The equations are based on

234 Transient Effects

TABLE 9.1
Specific Heat and Density of Materials Commonly Used in Electronic Equipment

Material	Specific Heat $\left(\dfrac{\text{watt-minute}}{\text{pound-}^\circ\text{C}}\right)$	Density (pound/inch³)
Aluminum	7.0	.10
Brass	3.0	.31
Copper	3.0	.32
Iron	3.4	.27
Lead	1.0	.41
Magnesium	7.9	.06
Steel	3.8	.29
Alumina	8.6	.14
Beryllia	6.4	.11
Epoxy	14.0	.04
Water	32.0	.04
FC-75	8.0	.06

the assumption that the temperature difference required to conduct heat from the component to the cooling fin is negligible compared to the temperature difference between the cooling fins and the surroundings. For most cases, this assumption is reasonable, and the error involved in the use of Equations 9.1 through 9.3 is small.

9.3 EXAMPLES

Transient thermal effects in electronic equipment can be calculated from the equations of this chapter. The correct use of this design information is illustrated in the section by the following three examples:

- The radiation and natural convection cooled heat sink described in Section 3.1 and Section 9.1.
- The forced air cooled heat sink described in Section 4.1.
- The forced liquid cooled heat sink described in Section 5.1.

The steady state temperature of these three heat sinks has been calculated in previous chapters.

EXAMPLE 1
An Aluminum Heat Sink Cooled by Radiation and Natural Convection

The warmup characteristics of a typical aluminum heat sink cooled by radiation and natural convection were discussed in Section 9.1. The heat sink was shown in Figure 9.1. The steady state characteristics of this heat sink were calculated in Example 1 of Section 3.4, and its transient characteristics will be calculated in this example.

The important dimensions of this heat sink were shown in Figure 3.9. As discussed in Chapter 3, when the component mounted on the heat sink is generating 60 W and the temperature of the surroundings is 20°C, the temperature difference between the heat sink and the surroundings is 80°C.

$$\frac{\Delta T}{Q} \text{ is, therefore, } \frac{80°C}{60 \text{ W}} = 1.33°C/W.$$

The first calculation will be the warm up characteristics of the heat sink. The temperature of the heat sink at any time after turnon is calculated from Equation 9.1:

$$\Delta T(\gamma) = \Delta T(\infty) - [\Delta T(\infty) - \Delta T(0)]e^{-\gamma/[(\Delta T/Q)cM]}.$$
$$= 80°C - [80°C - 0°C]e^{-\gamma/[1.33 °C/W \times 7.0 (W\text{-}min)/(lb\text{-}°C) \times .67 lb]}$$

where:

$\Delta T(\infty)$ is the final steady state temperature difference between the heat sink and the surroundings and is 80°C.

$\Delta T(0)$ is the initial temperature difference between the heat sink and the surroundings and is 0°C.

$\Delta T/Q$ is 1.33°C/W.

c is 7.0 (W-min)/(lb-°C) from Table 9.1.

M is .67 lb.

The weight M of the heat sink is calculated from the volume of the heat sink and the density of aluminum given in Table 9.1. The volume of the 16 fins, which are each .10 in. × .90 in. × 3.0 in., is 4.3 in³. The volume of the base of the heat sink is .20 in. × 4.0 in. × 3.0 in. or 2.4 in³. The total volume of the heat sink is 6.7 in³. From Table 9.1, the density of aluminum is .10 lb/in³, so the weight of the heat sink is .67 lb.

The calculation using Equation 9.1 is shown at several times in the following table, and these results were shown graphically in Figure 9.2.

Time (minutes)	$\gamma/[(\Delta T/Q)cM]$	$e^{-\gamma/[(\Delta T/Q)cM]}$	$\Delta T(\gamma)$ (°C)
5	.80	.449	40
10	1.60	.202	64
15	2.40	.091	73
20	3.20	.041	77

A particularly interesting time is when the temperature of the heat sink has come within 5% of its steady state value. From Equation 9.3:

$$(\gamma)_{5\%}^{-} = 3\left(\frac{\Delta T}{Q}\right)cM$$

$$= 3 \times 1.33°\text{C/W} \times 7.0 \text{ (W-min)/(lb-°C)} \times .67 \text{ lb}$$

$$= 19 \text{ min.}$$

The second calculation will be for the transient case when the heat sink is initially at 100°C with the electric component generating 60 W, and the power is suddenly reduced to 30 W. As shown in Chapter 3, with a dissipated power of 30 W, the temperature of the heat sink is 66°C when the surroundings are at 20°C. The temperature of the heat sink at any time after the power is reduced is calculated from Equation 9.1:

$$\Delta T(\gamma) = \Delta(\infty) - [\Delta T(\infty) - \Delta T(0)]e^{-\gamma/[(\Delta T/Q)cM]}$$

$$= 46°\text{C} - [46°\text{C} - 80°\text{C}]e^{-\gamma/[1.33°\text{C/W} \times 7.0 \text{ (W-min)/(lb-°C)} \times .67\text{lb}]}$$

where:

$\Delta T(\infty)$ is the final steady state temperature difference between the heat sink and the surroundings and is 46°C.

$\Delta T(0)$ is the initial temperature difference between the heat sink and the surroundings and is 80°C.

$\Delta T/Q$ is 1.33°C/W.

c is 7.0 (W-min)/(lb-°C).

M is .67 lb.

The calculation is shown at several times in the following table, and these results were shown graphically in Figure 9.3.

Time (minutes)	$\gamma/(\Delta T/Q)cM$	$e^{-\gamma/[(\Delta T/Q)cM]}$	$\Delta T(\gamma)$ (°C)
5	.80	.449	61
10	1.60	.202	53
15	2.40	.091	49
20	3.20	.041	47

FIGURE 9.5
Forced air cooled heat sink (Photo courtesy of Wakefield
Engineering, Inc.)

EXAMPLE 2
A Forced Air Cooled Copper Heat Sink

Figure 9.5 shows a photograph of a forced air cooled copper heat sink. The
steady state thermal characteristics of this heat sink were described in
Section 4.1 of Chapter 4 and were calculated in Example 1 of that chapter.

The heat transfer capability of this air cooled heat sink depends on the air
flow rate. At an air flow of 45 CFM, the temperature difference between the
heat sink and the inlet air required to transfer a given amount of power is
.25°C/W.

In order to compare this air cooled copper heat sink to the radiation and
natural convection cooled heat sink of Example 1, the time required for the
heat sink to come within 5% of its steady state value will be calculated, using
Equation 9.3:

$$(\gamma)_{5\%} = 3\left(\frac{\Delta T}{Q}\right)cM$$

$$= 3 \times .25°\text{C/W} \times 3.0 \text{ (W-min)/(lb-°C)} \times .29 \text{ lb}$$

$$= .65 \text{ min}$$

where:

$\Delta T/Q$ is .25°C/W.
c is 3.0 (W-min)/(lb-°C) from Table 9.1.
M is .29 lb.

The weight of the copper heat sink is calculated from its volume and density. The dimensions of the heat sink were given in Example 1 of Chapter 3. The volume of the base of the heat sink is 1 in. × 1.5 in. × 1/3 in. or .5 in³. The volume of the 13 fins is 13 × 1 in. × 1 in. × .030 in. or .4 in³. Therefore the total volume of the heat sink is .9 in³. The density of copper from Table 9.1 is .32 lb/in³, so the total weight of the heat sink is .29 lb.

This air cooled copper heat sink comes to within 5% of its steady state value in less than 1 min. In contrast, the heat sink of Example 1 required 19 min/.65 min = 29 times longer. The reasons for the faster warmup time of the air cooled copper heat sink are its lower weight, its greater heat transfer capability, and the lower specific heat of copper as compared to aluminum.

EXAMPLE 3
A Water Cooled Aluminum Heat Sink

Figure 9.6 shows a schematic diagram of a water cooled aluminum heat sink with the required cooling water reservoir and air-to-liquid heat exchanger. The steady state cooling characteristics of this heat sink were described in Section 5.1 of Chapter 5 and were calculated in Example 1 of that chapter.

The warmup characteristics of this electronic equipment depend both on the warmup characteristics of the heat sink and the warmup characteristics of the water in the cooling reservoir.

The first step is to calculate the time required for the heat sink to come within 5% of its steady state temperature. From Equation 9.3:

$$(\gamma)_{5\%} = 3\left(\frac{\Delta T}{Q}\right)cM$$

$$= 3 \times .01°\text{C/W} \times 7.0 \text{ (W-min)/(lb-°C)} \times .6 \text{ lb}$$

$$= .13 \text{ min}$$

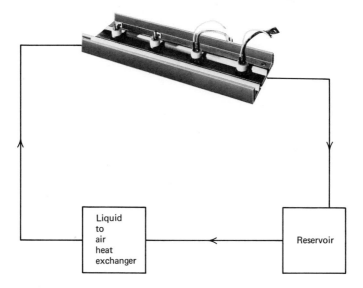

FIGURE 9.6
Schematic drawing of water cooled aluminum heat sink

where:

$\Delta T/Q$ is the temperature rise of the heat sink above the coolant required to transfer a given a mount of power and, as calculated in Example 1 of Chapter 5, is .01°C/W.

c is the specific heat of aluminum and is 7.0 (W-min)/(lb-°C) from Table 9.1.

M is the weight of the aluminum heat sink and is .6 lb.

The volume of the heat sink is 2 in. × 12 in. × 1/4 in., or 6 in³. The density of aluminum is .1 lb/in³ so the weight of the heat sink is .6 lb.

The next step is calculate the time required for the water to come to within 5% of its steady state temperature. The water is initially at the same temperature as the surroundings but it must increase above this temperature in order to transfer the heat to the surroundings in the liquid-to-air heat exchanger. The system will be assumed to hold 1 gal of water (231 in³) in the reservoir, heat exchanger, and cooling lines. The rate of change of water temperature per unit of power transferred in the liquid-to-air heat exchanger depends on the heat exchanger design, but will be assumed to be .01°C/W.

From Equation 9.3, the time required for the water to come within 5% of its steady state temperature is:

$$(\gamma)_{5\%} = 3\left(\frac{\Delta T}{Q}\right)cM$$

$$= 3 \times .01°C/W \times 32 \text{ (W-min)/(lb-°C)} \times 9.2 \text{ lb}$$

$$= 8.8 \text{ min}$$

where:

$\Delta T/Q$ is the temperature rise of the water required to transfer a given amount of power in the heat exchanger and is .01°C/W.

c is the specific heat of water and is 32 (W-min)/(lb-°C) from Table 9.1.

M is the weight of the water and is 231 in^3 × .04 lb/in^3 = 9.2 lb.

A time of 8.8 min is required for the temperature difference between the water and the surroundings to come within 5% of its steady state value. Therefore, the time required for the water temperature to stabilize in this cooling system is many times greater than the time required for the heat sink to stabilize in temperature. Consequently, the transient characteristics of the cooling water temperature is the significant factor in this cooling design.

9.4   REFERENCE

A useful reference on transient effects in electronic equipment is as follows:

1. Krauss, A. D., *Cooling Electronic Equipment*, Prentice Hall, Inc., Englewood Cliffs, N.J., 1965, pp. 12–29.

10

Thermal Measurement Techniques for Electronic Equipment

Previous chapters have discussed the cooling design of electronic equipment. The design information presented in those chapters gives only approximate answers. Therefore, after the cooling has been designed and the electronic equipment built, the thermal performance of the equipment must be measured.

Techniques for making the necessary thermal measurements on electronic equipment are presented in this chapter.

The thermal measurements that must be made on electronic equipment include:

1. Temperature of the electronic component, the heat sink, and the cooling fins.
2. Temperature distribution through the various parts.
3. Air flow rate.
4. Air pressure.
5. Liquid flow rate.
6. Liquid pressure.

The measurement of temperature and temperature distribution is discussed in Section 10.1. Techniques for measuring air flow rate and air pressure are presented in Section 10.2. The measurement of liquid flow rate and liquid pressure are described in Section 10.3.

One of the best ways of developing an adequate cooling design for electronic equipment is to build a thermal mockup in which the electronic components are replaced by dummy heat loads. In this way, the cooling can be designed and experimentally evaluated without fear of damage to the electronic components. Techniques for making thermal mockups of electronic equipment are described in Section 10.4.

10.1 MEASUREMENT OF TEMPERATURE AND TEMPERATURE DISTRIBUTION

The purpose of cooling electronic equipment is to keep the temperature of the electronic components below their safe operating temperature. Therefore, the measurement of the temperature of the electronic components, the heat sinks, the cooling fins, and the temperature distribution throughout these parts is one of the most important measurements that must be made on electronic equipment.

Useful equipment for measuring temperature and temperature distribution in electronic equipment include:

1. Thermocouples.
2. Thermistors.
3. Thermometers.
4. Temperature sensitive crayons, paints, and labels.
5. Liquid crystals.
6. Infrared detectors.

Typical examples of each of these temperature measuring equipments, their ranges of usefulness, and their proper use are discussed in this section.

Thermocouples

A thermocouple consists of two wires of unlike metals or alloys which are joined together at one end. When the junction where the wires are joined together is heated, a small voltage appears at the open ends of the wires. This small voltage depends on the types of materials used in the thermocouple and is proportional to the temperature difference between the junction and the open ends of the wires. Table 10.1 shows the four thermocouple types most commonly used for measuring temperatures in electronic equipment and compares their useful temperature ranges and the voltages developed when the junction is at 100°C and the open ends are at 0°C.

The following three requirements must be considered in using thermocouples to measure temperature:

1. Accurate measurement of the voltage across the open ends of the wires.
2. Control of the reference temperature at the open ends of the wires.
3. Compensation for the nonlinearity of the voltage-temperature characteristic of the thermocouple wire pair.

TABLE 10.1
Thermocouple Characteristics

Thermo-couple Type	Materials	Application	Useful Temperature Range (°C)	Voltage Developed*
T	Copper-constantan	Low tempera-ture	−260° to 350°	4.28
J	Iron-constantan	Reducing atmosphere	−150° to 750°	5.27
K	Chromel-alumel	Oxidizing atmosphere	−250° to 1250°	4.10
R	Alloy of 87% platinum and 13% rhodium	High tempera-ture	0° to 1500°	0.65

* When junction is at 100°C and open ends are at 0°C (millivolts).

The most accurate temperature measurements are made by using a voltage measuring device such as a potentiometer which draws no current and by immersing the open reference ends of the wires in an ice bath to maintain their temperature at 0°C.

For most temperature measurements on electronic equipment, an accuracy of ±2°C is satisfactory, and such elaborate procedures as an ice bath reference are not necessary.

A simple thermocouple meter which is commonly used for measuring temperature in electronic equipment is shown in Figure 10.1. The voltage generated by the thermocouple is used directly to operate a millivolt meter. The meter face is calibrated directly in degrees, and this calibration includes the nonlinear voltage-temperature relationship. A bimetal spring in the meter movement changes the pointer reference position to account for ambient temperature changes at the open reference ends of the thermocouple wires. Because the meter movement is determined by electric current, the resistance of the thermocouple wires must be controlled by selecting the proper wire length and diameter. The thermocouple meter shown in Figure 10.1 covers the temperature range from −15°C to 150°C, with an accuracy of 2%. It requires no batteries or external power supplies for its operation. The instrument is shown with the thermocouple wires mounted in a probe. The junction is at the tip of the probe, and the probe can be moved from point

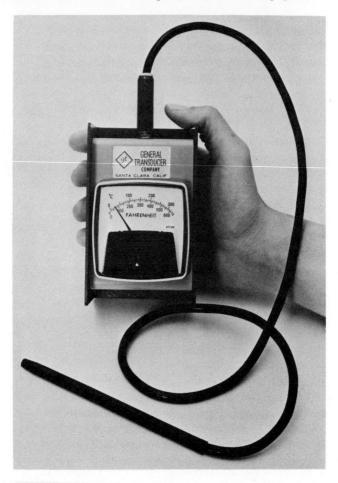

FIGURE 10.1
Thermocouple meter with thermocouple probe (Photo courtesy
of General Transducer Company)

to point in the electronic equipment to measure both temperature and
temperature distribution. The complete probe and meter costs from $75
to $100.

A more sophisticated thermocouple meter is shown in Figure 10.2. The
voltage from the thermocouple is amplified electronically, corrections for the
reference temperature and the nonlinearity of the voltage-temperature
characteristics of the wire pair are made electronically to the signal, and the
temperature is presented as a digital display. The instrument shown can
measure temperature over any range from −250°C to 1500°C with the
proper choice of thermocouple wires with an accuracy of .5°C. A digital

FIGURE 10.2
Digital thermocouple meter (Photo courtesy of Newport Laboratories, Inc.)

thermocouple meter of the type shown in Figure 10.2 costs in the $500 to $750 range.

The most important precaution to be taken in using thermocouples is to be sure that the thermocouple junction is in intimate thermal contact with the part whose temperature is to be measured. If the junction does not contact the part, the thermocouple will be measuring the temperature of the surrounding air and not the part temperature. Figure 10.3 shows correct and incorrect thermocouple mounting techniques. The best way to ensure good thermal contact is to spot weld or solder the thermocouple junction to the part, as shown in Figure 10.3A. If this cannot be done, the thermocouple junction must be wedged tightly against the part. As shown in Figure 10.3B, a small hole can be drilled in the part whose temperature is to be measured and thermocouple junction wedged against the inner diameter of the hole by a piano wire. Figure 10.3C shows incorrect mounting, where the piano wire forces the thermocouple wires against the inner diameter of the hole, but the junction does not touch the part and, so instead, measures the air temperature inside the hole.

When the thermocouple is mounted in the tip of a probe, such as with the thermocouple meter shown in Figure 10.1, the probe tip must be pressed tightly against the part whose temperature is being measured. Figure 10.3D shows the best technique for use with a thermocouple probe. The tip of the probe is first dipped in heat sink compound. As discussed in Chapter 2, the use of heat sink compound significantly reduces the thermal interface between the part and the thermocouple junction.

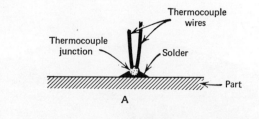

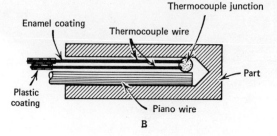

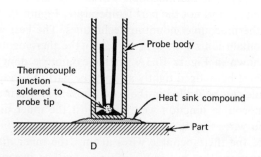

FIGURE 10.3
Techniques for mounting thermocouples on electronic equipment. A. Thermocouple soldered to part. B. Correct mounting without soldering. C. Incorrect mounting. D. Thermocouple in probe

Thermistors

Thermistors are electronic devices whose electrical resistance changes rapidly with temperature. Therefore, they are useful as temperature sensors. Thermistors have much greater thermal sensitivity than thermocouples, and do not require a constant temperature reference like thermocouples. The useful temperature range of thermistors is limited, usually from −80°C to 150°C. In contrast, as shown by Table 10.1, thermocouples can cover a much wider temperature range from −260°C to 1500°C. However, the normal operating range of electronic equipment is limited also, so thermistors are extremely useful for temperature measurements in electronic equipment.

FIGURE 10.4
Thermistor temperature meter (Photo courtesy of Yellow Springs Instrument Company)

Figure 10.4 shows a typical thermistor temperature meter, and Figure 10.5 shows the variety of thermistor probes that can be used with it. The temperature meter with any of the probes shown covers the temperature range from −40°C to 150°C with .5°C accuracy. The probes shown in the right-hand side of Figure 10.5 are most useful for measurements in electronic equipment. The tip of the fourth probe from the right, for example, is only 3/16 in. in diameter by 1/16 in. high. It can be mounted directly on almost any electronic component.

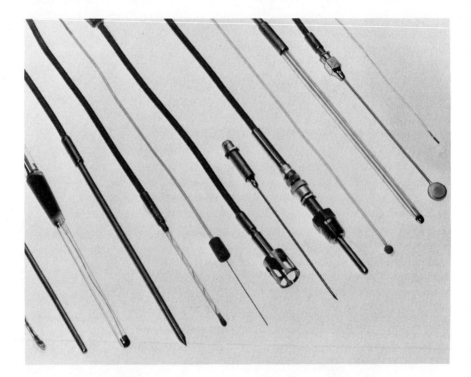

FIGURE 10.5
Thermistor temperature probes for use with thermistor temperature meter shown in Figure 10.4 (Photo courtesy of Yellow Springs Instrument Company)

The same precaution that is required with thermocouples of insuring that the sensing end of the probe is in good thermal contact with the part whose temperature is being measured must also be observed with thermistors. Correct methods of mounting the thermistor probe on the electronic component are shown in Figures 10.3B and 10.3D.

Thermometers

Thermometers are most commonly used for measuring the temperature of the cooling air or coolant liquid in electronic equipment. In order to accurately measure temperature, the bulb of the thermometer must be in intimate contact with the part whose temperature is to be measured. It is easy to provide intimate contact between the thermometer bulb and the cooling air or coolant liquid, but difficult to provide good contact between the thermometer bulb and the electronic component itself.

Temperature Sensitive Crayons, Paints, and Labels

Temperature sensitive crayons, paints, and labels melt or change color at specific temperatures. These temperature indicators come in sets and each member of the set melts or changes its characteristics at a different temperature.

Figure 10.6 shows a temperature sensitive crayon. One hundred different crayons are available with calibrated melting temperatures from 38°C to 1800°C. At the low temperature end of this range, the crayons of the set have melting temperatures 2°C apart.

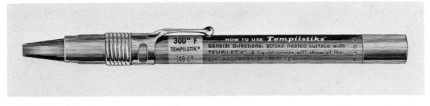

FIGURE 10.6
Temperature sensitive crayon (Photo courtesy of Tempil Corporation)

The crayons each have different colors and change their colors somewhat with temperature. However, their color change is in no way related to temperature. The crayons indicate temperature by melting when their particular rated temperature is reached within a tolerance of $\pm 1\%$ of their rating. For most applications in electronic equipment, the electronic component is marked with the crayon before the equipment is turned on. When the component reaches the specific temperature of the crayon, the mark melts. By marking a component with an array of lines, each from a different crayon, a temperature distribution can be obtained.

In some applications, the simple method of marking the part with the crayon may not be satisfactory because the mark evaporates after prolonged heating or is gradually absorbed into the surface so that too little material is left for unambiguous observation. In this case, the part may be touched or stroked with the crayon at regular intervals. The crayon will leave a dry mark at temperatures below its rating and a liquid streak at temperatures above its rating.

The temperature sensitive material in the crayon is also available mixed in an inert, volatile, nonflammable vehicle. The resulting paint can be brushed or sprayed onto the electronic component being tested. When the component reaches a specific temperature, the paint will melt. By watching the regions over the part where melting occurs as the power dissipation is slowly increased, a temperature distribution can be obtained.

Like the temperature sensitive crayons, the temperature sensitive paints have different colors and change their color on heating, but this color change is never to be interpreted as a temperature indication. The temperature sensitive paints come in 100 different temperature ratings over the temperature range from 38°C to 1800°C. Like the crayons, the specific melting temperature of each member of the set is only 2°C different at the low end of the range.

A certain amount of skill is necessary in interpreting exactly when the temperature sensitive crayons or paints actually melt. A certain amount of skill is also needed in applying just the right amount of paint with the proper consistency. The best way to obtain this skill is to test the crayons and paint on sample pieces whose temperature is simultaneously monitored by a thermocouple. Once their unique behavior has been learned, the crayons or paints form a simple, inexpensive, and versatile means of making temperature measurements. A crayon or a small bottle of paint costs about $3 and is sufficient for many temperature measurements.

The major disadvantage of the temperature sensitive paints and crayons is that the melting is irreversible, so that for repeated measurements the part must be cleaned and remarked. Also the component must be observable during operation.

Figure 10.7 shows a temperature sensitive label mounted on a high power transmitter tube. The labels can be made to contain from one to eight temperature indicators in any standard calibrated temperature increments from 37°C to 593°C. When exposed to the rated temperature, the indicator turns from pastel to black. The change in color is permanent and irreversible. The reaction occurs in less than 1 sec and within $\pm 1\%$ of the rated temperature.

The label on the high power transmitter tube shown in Figure 10.7 indicates that the tube envelope has reached a temperature of 250°F, as indicated by the first two of the four temperature sensitive indicators having turned black, but it has not reached the higher temperature of 300°F since the other two indicators have not changed color.

Temperature sensitive labels provide a more positive indication that a given temperature has been reached than do the temperature sensitive crayons or

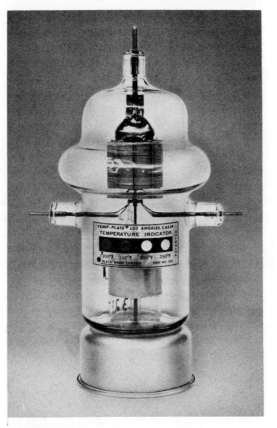

FIGURE 10.7
Temperature sensitive labels mounted on a high
power transmitter tube (Photo courtesy of William
Wahl Corporation)

paints. Because the indication is permanent, the electronic equipment need
not be observed during operation.

The major disadvantage of the temperature sensitive labels is their high
cost. The labels cost approximately $1 per label and can be used only once.

Liquid Crystals

"Liquid crystals" are cholesterol compounds which change their color
through the entire visible spectrum as their temperature changes. The
material can be painted on an electronic component and, as the component
temperature changes, the liquid crystal paint will change from colorless to
red, to orange, to yellow, to green, to blue, to violet, and to colorless again.

The temperature at which a specific color occurs and the temperature range from the red to the violet ends of the spectrum can be varied by the proper choice of liquid crystal formulation. The color change is completely reversible, so that the measurement can be repeated as often as desired.

Typical variation of the color of liquid crystal paints as a function of temperature is shown in Figure 10.8. Depending on the exact formulation, the paint can change through the entire visible spectrum from red to violet in a range of 3°C centered around 25°C, 45°C, 65°C, as shown in Figure 10.8, or around any intermediate temperature range. Alternately, the temperature response of the material can be broadened so that the material changes color gradually across the visible spectrum from red to violet over a 40°C temperature range. With this latter composition, liquid crystal paints are particularly useful for measuring temperature distributions in electronic equipment. The hottest parts which are near 70°C will be blue and violet in temperature, whereas the coolest parts in the range just above room temperature will be yellow and red. Midway between the two temperatures extremes at around 50°C, the parts will be green in color.

A typical liquid crystal kit for making temperature measurements, such as supplied by the Vari-Light Corporation of Cincinnati, Ohio consists of three bottles of liquid crystals. By mixing the proper proportions of two of the bottles, the temperature at which the liquid crystal changes color across the entire visible spectrum can be moved to any temperature from 25°C to 75°C. By adding appropriate amounts of the third bottle, the temperature response can be broadened from 3°C to 40°C. Temperature resolutions to within 2°C

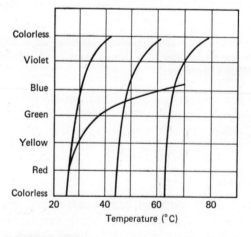

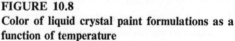

FIGURE 10.8
Color of liquid crystal paint formulations as a function of temperature

can be made at spacings as close as .005 in. Response time is less than 1 sec. The liquid crystal colors appear best against a black background. Consequently, the liquid crystal kit contains a bottle of compatible black water base paint. For optimum results, the electronic components must first be painted with the black paint and then the liquid crystal paint applied. Similar kits are available to cover the temperature range from 80°C to 135°C, and from 135°C to 180°C.

Liquid crystals are inexpensive. A complete kit such as described above costs $50 and provides material for many measurements.

Infrared Detectors

Infrared detectors are able to measure electronic component temperatures without making contact with the component. Temperature is measured by the infrared thermal radiation from the component, which is a function of component's surface temperature.

Figure 10.9 shows one type of noncontacting infrared thermometer measuring the temperature of a transistor case. This instrument can measure

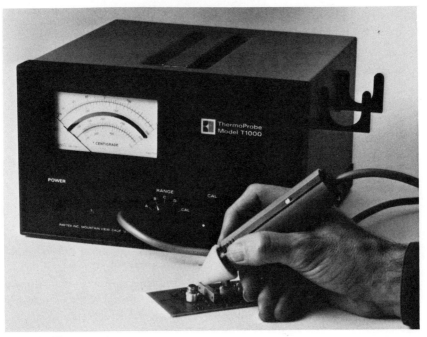

FIGURE 10.9
Infrared thermometer measuring temperature of an electronic component (Photo courtesy of Raytek)

FIGURE 10.10
Infrared scanning microscope (Photo courtesy of AGA Corporation)

areas as small as .050 in. in diameter when the probe is 1 in. from the surface. Temperatures from 20°C to 500°C can be measured in four switch selectable subranges. An infrared detector of this type costs $1750.

One of the most sensitive and versatile instruments for measuring temperature distributions in electronic equipment is the infrared detector shown in Figure 10.10. This instrument consists of an infrared microscope and mechanical scanning system of infrared prisms which focus the heat from a small area of the electronic component onto the sensitive infrared detector in the instrument. To obtain the required sensitivity, the infrared detector must be refrigerated in a liquid nitrogen bath which is contained in the cylindrical

Microcircuit viewed through the
visible light optics.

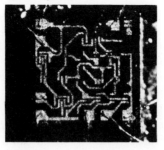

A.

Same microcircuit scanned
through the IR optics and displayed as a thermal image.

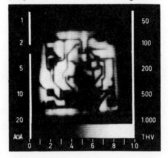

B.

FIGURE 10.11
Microcircuit and its thermal
image (Photo courtesy of AGA
Corporation)

tank around the instrument. The output of the infrared detector is a voltage
whose magnitude is proportional to the temperature at the particular spot on
the component. This electrical signal is processed and is displayed as intensity
modulation on an oscilloscope.

The infrared scanning system moves the spot where temperature is measured
across the surface of the component and electrical voltages proportional to
the spot location are applied to the oscilloscope horizontal and vertical axes.
In this way, a thermal image of the electronic component is presented on the
oscilloscope.

Figure 10.11A shows an electronic microcircuit as viewed through the
optical microscope located on the front of the infrared detector instrument.

The field of view is approximately 1/4 in. × 1/4 in. Figure 10.11B shows the thermal image of this microcircuit as measured by the infrared detector and displayed on the oscilloscope. The sensitivity is adjusted so that a 10°C temperature difference appears between white and black. The thermal image can also be displayed on a color cathode ray tube and each degree of the 10°C temperature difference can be shown as a different color.

Since the temperature information is taken from the infrared detector as an electrical signal, it can be processed in various ways for displaying the information. The temperature range displayed can be varied from a few degrees to a few hundred degrees. The signal can also be processed so that single isotherms are displayed or a three-dimensional temperature profile can be presented.

The instrument shown in Figure 10.10 can measure extremely small temperature differences, can operate over wide temperature ranges, and can provide excellent dimensional resolution. For example, with infrared optics that provide a 1/4 in. × 1/4 in. field of view, temperature differences as small as .1°C can be resolved at .004 in. spacings. With different optics which restrict the field of view to .050 in. × .050 in., the resolving power is .0004 in. for temperature differences of 2.5°C.

The instrument can measure over any temperature range from −30°C to 850°C.

The disadvantage of the infrared detector shown in Figure 10.10 is its high cost of approximately $25,000, which includes the signal processing and display equipment.

10.2   MEASUREMENT OF AIR FLOW AND AIR PRESSURE

A schematic drawing of the thermal measurement setup for a forced air cooled electronic component is shown in Figure 10.12.

Thermometers are shown being used to measure inlet and outlet air temperature. Thermocouples or thermistors could also be used for this measurement. A thermocouple is shown being used to measure the temperature of the component, and the thermocouple is welded to the component to ensure a good thermal contact.

Air pressure is measured with a water tube manometer. Pressure is expressed in inches of water, which is the height of a column of water that the pressure will support.

To measure static pressure, the tubing from the manometer must be oriented at right angles to the direction of air flow, as shown in Figure 10.12.

The two arms of the pressure manometer are connected to opposite sides of component and so measure the pressure difference required to force air through the component's cooling fins.

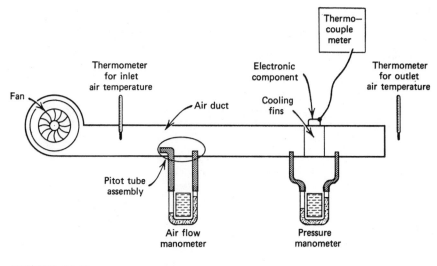

FIGURE 10.12
Schematic diagram of a thermal measurement set up for a forced air cooled electronic component

To increase reading accuracy, the manometer tube can be inclined. The air pressure will raise the column of water a certain vertical height. By inclining the tube, the water column length can be greatly increased so that accurate measurements can be made to a fraction of an inch. To further increase sensitivity, the tube can be filled with an oil which has a lower density than water. Figure 10.13 shows a typical inclined manometer used for air pressure measurements. The instrument scale covers a water pressure range of 0 to 6 in.

Air flow rate is determined by measuring the velocity of the air, and then multiplying the air velocity by the cross-sectional area of the duct, as shown by the following formula:

$$f = .007vA \qquad (10.1)$$

where:

f is total air flow rate (CFM).
v is air velocity in feet per minute.
A is cross sectional area of air duct in inch^2.

Air velocity can best be measured with a pitot tube at high flow rates and with a hot wire anemometer at low flow rates. A pitot tube assembly is shown in the schematic measurement setup of Figure 10.12.

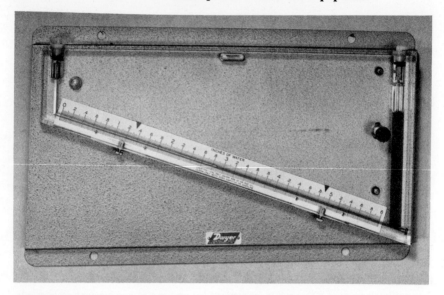

FIGURE 10.13
Inclined manometer (Photo courtesy of F. W. Dwyer Mfg. Company)

The pitot tube assembly consists of two probes. The probe with its opening pointing into the direction of air flow measures both the air velocity pressure and the static pressure. The probe with its opening at right angles to the flow measures only the static pressure. The difference between these two pressures is determined directly by connecting the probes to the opposite arms of a water column manometer. As with air pressure measurements, an inclined manometer tube is usually used to increase sensitivity. Air velocity can be calculated from the following formula:

$$v = 4000\sqrt{\beta} \qquad (10.2)$$

where:

v is the air velocity in feet per minute.
β is the height of the manometer water column in inches.

The constant term in Equation 10.2 applies to the case of sea level air at a temperature of 25°C. If measurements are made with air at other temperatures, the constant multiplying factor of the equation is modified as shown in Figure 10.14.

Complete pitot tube assemblies for measuring air velocity are shown in Figure 10.15. Each of the assemblies shown are only 1/8 in. in diameter and contain both of the two probes shown schematically in Figure 10.12. The

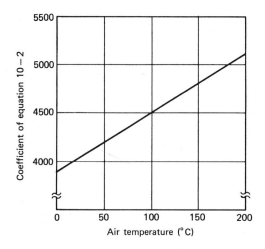

FIGURE 10.14
Coefficient of Equation 10.2 as a function of air temperature

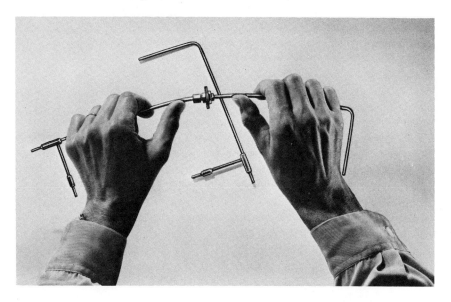

FIGURE 10.15
Pitot tube assemblies (Photo courtesy of F. W. Dwyer Mfg. Company)

two probes are mounted coaxially. The inner probe has its opening directed into the direction of air flow and measures total pressure. The outer probe has several small openings around its diameter to measure static pressure. Connections are made from these two coaxial probes at the lower end of the assembly to a water column manometer.

A typical example will illustrate the use of Equations 10.1 and 10.2. Assume the air duct has a cross section of 3 in. × 3 in., and the measured height of the water column in the manometer which is connected to the pitot tube assembly is .16 in. From Equation 10.2, the air velocity is $4000 \times \sqrt{.16}$ = 1600 ft/sec. From Equation 10.1, the air flow rate is .007 × 1600 × 3 × 3 = 100 CFM.

At low air flow rates, the water column height in even an inclined manometer is difficult to determine, and air velocity is more accurately measured with a hot wire anemometer. Such an instrument is shown in Figure 10.16. It measures air velocity up to 6000 ft/min with ± 2% accuracy in three switch selectable ranges. The sensing probe is 3/8 in. in diameter and 9 in. long.

Air velocity almost always varies across the duct cross section. Therefore, when using either the pitot tube or the hot wire anemometer method, the air

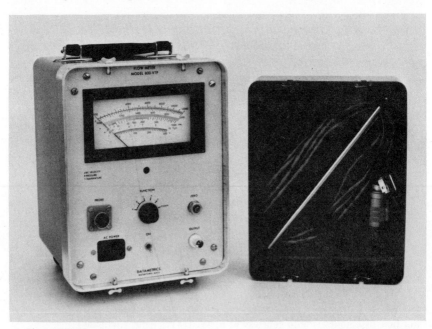

FIGURE 10.16
Hot wire anemometer air flow meter (Photo courtesy of Datametrics Division of C. G. S. Scientific Corporation)

velocity must be measured at several locations across the cross section and the results averaged.

A rough estimate of the air flow rate can be made by measuring the temperature rise of the air as it absorbs power from the electronic component and then calculating the flow rate from Equation 4.2 of Chapter 4. For example, if the power dissipated by the component is 400 W, and the temperature rise of the air is 7°C, then the air flow is $1.73/(\Delta T/Q) = 1.73/(7°C/400 \text{ W})$ = 100 CFM. The outlet air temperature will vary across the cross section of the duct, so temperature readings must be taken at several locations and the results averaged.

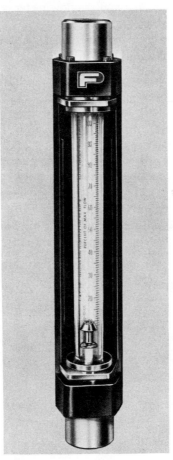

FIGURE 10.17
Liquid flow meter (Photo courtesy of Fischer and Porter Company)

10.3   MEASUREMENT OF LIQUID FLOW RATE AND LIQUID PRESSURE

Measurements of liquid flow and liquid pressure are much easier than the measurement of air flow and air pressure discussed in the previous section, because the liquid is always contained in a small hose or duct.

Liquid flow is normally measured with a flow meter of the type shown in Figure 10.17. The flow meter is connected in series in the coolant line. The coolant flowing against the float raises it in the flow meter column. The height of the float depends on:

1. The geometry of the float and the flow meter column.
2. The density of the float material.
3. The density and viscosity of the coolant.
4. The coolant flow rate.

The column and float dimensions can be adjusted by the flow meter manufacturer for any flow range and any liquid. A graduated scale is usually supplied with the flow meter and is calibrated directly in gallons per minute. Measurements are made by reading the location of the top of the float against the graduated scale.

The pressure at any point in a liquid cooling system or the pressure in a liquid filled electronics package is easily measured with a pressure gauge of the type shown in Figure 10.18. The pressure sensing element is a thin walled

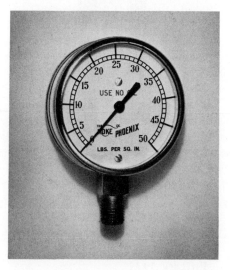

FIGURE 10.18
Pressure gauge (Photo courtesy of Hoke, Inc.)

tube, coiled into a circular arc. One end of the tube is sealed, and connected to the pointer. The opposite end of the tube is mounted to the gauge case and is connected to the pressure source. When pressure is applied, the free end of the tube is displaced and drives the pointer. Positive or negative pressures can be measured, relative to atmospheric. Typical accuracies are $\pm 1\%$.

10.4 THERMAL MOCKUPS

It is often useful to build a thermal mockup of an electronics equipment, so that its cooling performance can be evaluated over a wide range of operating conditions without damaging the electronic components under those conditions where the cooling is not sufficient. In the thermal mockup, the electronic components are replaced by heating elements which are operated from 60 cycle AC power.

Typical heating elements that can be used as dummy electronic components are shown in Figure 10.19. The element on the left is 3/8 in. in diameter by

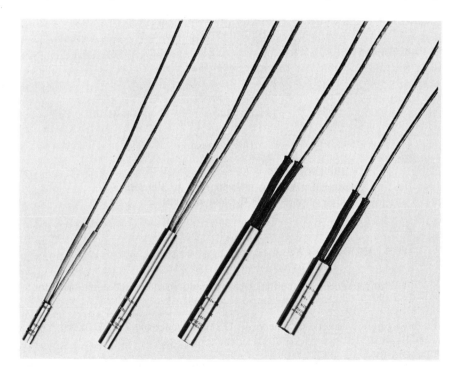

FIGURE 10.19
Heating elements used to simulate electronic components in thermal mockups (Photo courtesy of Watlow Electric Manufacturing Co.)

1 1/2 in. long and can supply up to 200 W of power. The larger elements can provide greater power.

The power supplied to these heating elements can be controlled by a variable auto transformer (Variac) so that cooling performance can be evaluated over a range of power dissipation. All cooling fins, heat sinks, and thermal joints should be made exactly as they will be in the final equipment. In this way, all possible thermal problems can be evaluated.

Figure 10.20 shows a heating element which simulates a high power SCR The resistive heating element is mounted inside a case which has the same external dimensions as an SCR. The copper hex-shaped base and stud are identical to the SCR design. This dummy load can be clamped to a heat sink, and all details of the heat transfer process simulated, including the thermal interface between the component and the heat sink. The experimental measurements of the effect of clamping torque and heat sink compound that were shown in Figure 2.11 of Chapter 2 were made with the dummy load shown in Figure 10.20.

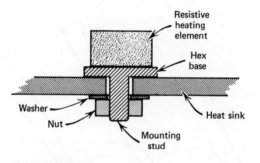

FIGURE 10.20
Resistive heating element used to simulate a high power SCR in thermal mockups

10.5 REFERENCES

Useful references on thermal measurements in electronic equipment are as follows:

1. *Temperature Measurement Handbook*, Omega Engineering, Inc., Stanford, Conn., 1971.

2. Goetter, D. F., and Anderson, W., " Standardized thermal testing—the number one way to evaluate semiconductor cooling," *EDN/EEE Magazine*, June 17, 1971, pp. 31–36.

3. Lauriente, M., and Fergason, J. L., "Liquid crystals plot the hot spots," *Electronic Design Magazine*, **19**, Sept. 13, 1967, pp. 71–79.

4. "New generation of infrared devices makes non-destructive testing easier," *Product Engineering Magazine*, Aug. 3, 1970, pp. 24–25.

5. Dietz, H. G., *Forced Air Cooling Primer for the Electronics Engineer*, Harry G. Dietz Co., Long Island City, New York, 1964.

6. Goldman, W. E., "How to Measure Heat Sink Characteristics for High Power Applications," *EEE Magazine*, July, 1965, pp. 10–11.

Reference 1 describes thermocouple characteristics and typical thermocouple and thermistor temperature meters.

Reference 2 describes mounting of thermocouples on various standard semiconductor packages and the measurement of temperature characteristics of these components.

Reference 3 describes the use of liquid crystals to measure the temperature distribution in electronic equipment. This article shows interesting color photographs of the temperature distribution during operation of integrated circuits.

Reference 4 summarizes the use of infrared detectors for measuring temperature distributions in electronic equipment and describes available equipment.

Reference 5 describes the measurement of air pressure and air flow.

Reference 6 discusses the use of thermal mockups for evaluating semiconductor heat sink characteristics.

Appendices

APPENDIX A

Definition of Symbols

Symbol	Definition	Units
A	cross-sectional area of duct through which cooling liquid or cooling air flows	$inch^2$
b	cylinder length for radial heat flow (See Figure A)	inch
c	specific heat	$\dfrac{watt\text{-}minute}{pound\text{-}^\circ C}$
C	specific heat	$\dfrac{BTU}{pound\text{-}^\circ F}$
d	diameter of cylindrical component (See Figure B)	inch
d_i	inner diameter for radial heat flow (See Figure A)	inch
d_o	outer diameter for radial heat flow (See Figure A)	inch
F	total coolant flow rate	gallon/minute
f	total air flow rate	CFM
g	specific gravity of coolant relative to water	—
k	thermal conductivity	$\dfrac{watt}{inch\text{-}^\circ C}$
K	thermal conductivity	$\dfrac{BTU}{hour\text{-}foot^2 - {}^\circ F/foot}$

268

Symbol	Definition	Units
L	length of coolant ducts in direction of coolant flow	inch
M	weight of heat sink or weight of coolant	pound
n	number of cooling ducts	—
ΔP	coolant pressure drop required to force coolant through the duct	psi
Δp	air pressure drop	inches of water
Q	power	watts
Q_r	power transferred by radiation per unit area from a perfect, unshielded radiator	watts/inch^2
Q_c	power transferred per unit area by natural convection from 1 in. high vertical fin	watts/inch^2
R	reduction in effective radiating surface area due to shielding of adjacent surfaces	—
R_1	natural convection factor due to surface orientation and geometry	—
R_2	natural convection factor due to altitude	—
S	surface area from which heat is transferred	inch^2
T	temperature	°C
ΔT	temperature difference	°C
$\Delta T/Q$	steady state temperature difference required to conduct or transfer a unit amount of heat	°C/watt
v	air velocity	feet/minute
w	width of cooling ducts (See Figure C)	inch
y	thickness of cooling fins (See Figure C)	inch
z	height of cooling fins above base (See Figure C)	inch
α	cross sectional area of material through which heat is conducted (See Figure D)	inch^2

Symbol	Definition	Units
β	height of water manometer column	inch
γ	time	minutes
λ	length through which heat is conducted (See Figure D)	inch
ϕ	dimensionless duct geometry factor (equal to 4π times the cross-sectional area of the duct divided by the square of its perimeter)	—
θ	thermal resistance	°C/watt
ε	radiation emissivity	—
μ	viscosity of coolant	$\dfrac{\text{pound}}{\text{hour-foot}}$

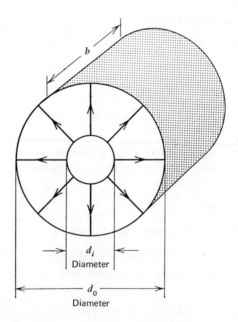

FIGURE A

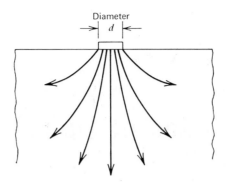

FIGURE B

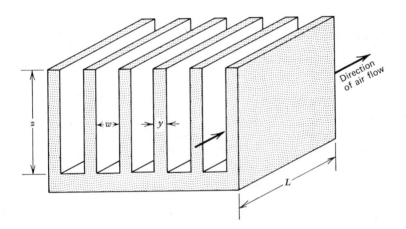

FIGURE C

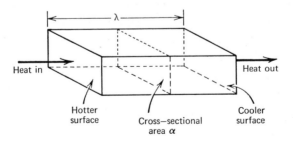

FIGURE D

APPENDIX B

Conversion of Units

The units of measurement used throughout the text and in all formulas have been chosen to be the easiest for electronic equipment designers to use. A variety of other systems of units are used in the listed references and other published data. Conversion factors are presented in this appendix to convert from these other units to the units used in this book.

For example, Table B-1 gives conversion factors for *power*. To convert power to watts from any of the units of power given in the table, multiply by the factor listed.

For example, assume power is given as 5 BTU/hr. To convert to watts, multiply the 5 BTU/hr by the conversion factor in the table of .293.

$$5 \text{ BTU/hr} \times .293 = 1.47 \text{ W}$$

1. POWER (watts)

UNITS	MULTIPLICATION FACTOR TO CONVERT TO WATTS
joule/second	1
gram-calorie/second	4.186
BTU/hour	.293
BTU/minute	17.6
Horsepower	746
Foot-pound/second	1.36

2. POWER DENSITY (watts/inch^2)

UNITS	MULTIPLICATION FACTOR TO CONVERT TO WATTS/INCH^2
$\dfrac{\text{watts}}{\text{centimeter}^2}$	6.45
$\dfrac{\text{watts}}{\text{meter}^2}$	6.45×10^{-4}
$\dfrac{\text{calorie/second}}{\text{centimeter}^2}$	27.0
$\dfrac{\text{BTU}}{\text{hour-foot}^2}$	2.035×10^{-3}
$\dfrac{\text{horsepower}}{\text{foot}^2}$	5.179

3. THERMAL CONDUCTIVITY $\left(\dfrac{\text{watts}}{\text{inch } ^\circ\text{C}}\right)$

UNITS	MULTIPLICATION FACTOR TO CONVERT TO WATT/INCH-$^\circ$C
$\dfrac{\text{gram-calorie}}{\text{second-centimeter}^2\text{-}^\circ\text{C/centimeter}}$	10.6
$\dfrac{\text{watt}}{\text{centimeter}^2\text{-}^\circ\text{C/centimeter}}$	2.54
$\dfrac{\text{BTU}}{\text{hour-foot}^2\text{-}^\circ\text{F/foot}}$	.044
$\dfrac{\text{BTU}}{\text{hour-foot}^2\text{-}^\circ\text{F/inch}}$	.0037

4. THERMAL CONDUCTIVITY $\left(\dfrac{\text{BTU}}{\text{hour-foot}^2\text{-°F/foot}}\right)$

UNITS	MULTIPLICATION FACTOR TO CONVERT TO $\dfrac{\text{BTU}}{\text{hour-foot}^2\text{-°F/foot}}$
$\dfrac{\text{watt}}{\text{inch °C}}$	22.8
$\dfrac{\text{watt}}{\text{centimeter}^2\text{-°C/centimeter}}$	58.0
$\dfrac{\text{gram-calorie}}{\text{second-centimeter}^2\text{-°C/centimeter}}$	242
$\dfrac{\text{BTU}}{\text{hour-foot}^2\text{-°F/inch}}$	.083

5. SPECIFIC HEAT $\left(\dfrac{\text{watt-minute}}{\text{pound-°C}}\right)$

UNITS	MULTIPLICATION FACTOR TO CONVERT TO $\dfrac{\text{WATT-MINUTE}}{\text{POUND-°C}}$
$\dfrac{\text{joule}}{\text{gram-°C}}$	7.5
$\dfrac{\text{calorie}}{\text{gram-°C}}$	31.5
$\dfrac{\text{BTU}}{\text{pound-°F}}$	31.5

6. SPECIFIC HEAT $\left(\dfrac{\text{BTU}}{\text{pound-}^\circ\text{F}}\right)$

UNITS	MULTIPLICATION FACTOR TO CONVERT TO $\dfrac{\text{BTU}}{\text{POUND-}^\circ\text{F}}$
$\dfrac{\text{joule}}{\text{gram-}^\circ\text{C}}$	.239
$\dfrac{\text{calorie}}{\text{gram-}^\circ\text{C}}$	1.00
$\dfrac{\text{watt-minute}}{\text{pound-}^\circ\text{C}}$	.032

7. COOLANT FLOW $\left(\dfrac{\text{gallon}}{\text{minute}}\right)$

UNITS	MULTIPLICATION FACTOR TO CONVERT TO GALLON/MINUTE
CFM	7.5
$\dfrac{\text{foot}^3}{\text{second}}$	450
$\dfrac{\text{meter}^3}{\text{minute}}$	264
$\dfrac{\text{liter}}{\text{second}}$	15.7
$\dfrac{\text{centimeter}^3}{\text{second}}$	.0157

8. AIRFLOW (CFM)

UNITS	MULTIPLICATION FACTOR TO CONVERT TO CFM
gallon/minute	.134
meter^3/minute	35.3
liter/second	2.1
centimeter^3-second	2.1×10^{-3}
foot^3/second	60

9. PRESSURE (psi)

UNITS	MULTIPLICATION FACTOR TO CONVERT TO PSI
inches of water	.036
feet of water	.434
inches of mercury	.491
millimeters of mercury	.019
kilogram/centimeter^2	14.2
atmospheres	14.7

10. AIR PRESSURE (inches of water)

UNITS	MULTIPLICATION FACTOR TO CONVERT TO INCHES OF WATER
psi	27.7
feet of water	12
inches of mercury	13.6
millimeters of mercury	.535
kilograms/centimeter^2	394
atmospheres	407

11. LENGTH (inches)

UNITS	MULTIPLICATION FACTOR TO CONVERT TO INCHES
feet	12
yards	36
centimeters	.394
meters	39.4
kilometers	3.94×10^4

12. AREA (inch^2)

UNITS	MULTIPLICATION FACTOR TO CONVERT TO INCH^2
feet^2	144
yards^2	1296
centimeter^2	.155
meter^2	1550

13. VOLUME (inch^3)

UNITS	MULTIPLICATION FACTOR TO CONVERT TO INCH^3
foot^3	1728
yard^3	4.67×10^4
gallons	231
centimeter^3	.061
liters	61.0
meter^3	6.10×10^4
fluid ounces	1.805

14. WEIGHT (pounds)

UNITS	MULTIPLICATION FACTOR TO CONVERT TO POUNDS
ounces	.063
grams	2.21×10^{-3}
kilograms	2.21

15. TIME (minutes)

UNITS	MULTIPLICATION FACTOR TO CONVERT TO SECONDS
seconds	.0167
hours	60

16. DENSITY $\left(\dfrac{\text{pounds}}{\text{inch}^3}\right)$

UNITS	MULTIPLICATION FACTOR TO CONVERT TO POUNDS/INCH³
pounds/foot³	5.79×10^{-4}
grams/centimeter³	.036
kilograms/meter³	3.6×10^{-5}
ounces/inch³	.0625

17. VISCOSITY $\left(\dfrac{\text{pound}}{\text{hour-foot}}\right)$

UNITS	MULITPLICATION FACTOR TO CONVERT TO $\dfrac{\text{POUND}}{\text{HOUR-FOOT}}$
centipoise	2.42
$\dfrac{\text{gram}}{\text{second-centimeter}}$	242
$\dfrac{\text{pound}}{\text{second-foot}}$	3600
$\dfrac{\text{kilogram}}{\text{hour-meter}}$	.68

NOTE: Viscosity is often given as kinematic viscosity, which is viscosity divided by the specific gravity of the liquid. The most common unit of kinematic viscosity is "centistokes," which is equal to the viscosity of the liquid in centipoise divided by its specific gravity.

18. TEMPERATURE (°C)

UNITS	CONVERSION FACTOR TO CONVERT TO °C
°F (Fahrenheit)	.556 (°F − 32)
°K (Kelvin)	(°K − 273)
°R (Rankine)	.556 (°R − 492)

A graphical conversion of F° to °C is shown in Figure E.

CONVERSION OF °F TO °C

FIGURE E

Index